Dr Alexis Willett is a science communicator who aims to make science accessible to all. She has a PhD in biomedical science from the University of Cambridge, where she studied at the Medical Research Council's Human Nutrition Research unit. She has taught human physiology and published on a wide range of health subjects. Alexis spends much of her time turning cutting-edge research and health policy jargon into something meaningful for the public, patients, doctors and policy makers. In her spare time, Alexis drinks a lot of rooibos tea.

Her first book, *How Much Brain Do We Really Need?*, with Jennifer Barnett, was published by Robinson in December 2017.

ALSO BY ALEXIS WILLETT

How Much Brain Do We Really Need?

Drinkology

....

*The Science of What We Drink and
What It Does to Us, from Milks to Martinis*

DR ALEXIS WILLETT

ROBINSON

ROBINSON

First published in Great Britain in 2019 by Robinson

Copyright © Alexis Willett, 2019

1 3 5 7 9 10 8 6 4 2

The moral right of the author has been asserted.

A CIP catalogue record for this book is available from the British Library

ISBN: 978-1-47214-247-4

Typeset in Scala by Hewer Text UK Ltd, Edinburgh
Printed and bound in Great Britain by Clays Ltd, Elcograf S.p.A.

Papers used by Robinson are from well-managed
forests and other responsible sources

Robinson
An imprint of
Little, Brown Book Group
Carmelite House
50 Victoria Embankment
London EC4Y 0DZ

An Hachette UK Company
www.hachette.co.uk

www.littlebrown.co.uk

To my family, with all their exacting drinks preferences.

Acknowledgements

I would like to thank all my friends and family who prompted the ideas and questions explored in this book. Without our many chance conversations, I may have ended up writing simply about water, tea and wine! You helped ferment a more rounded, full-bodied product. Thank you so much for your enduring enthusiasm and support (or at least pretending to be interested).

Thanks also to the team at Robinson for letting me have a second go at this book-writing business.

Naturally, any inaccuracies are entirely my own and I apologise in advance for all of them.

Contents

....

	Aperitif	xi
CHAPTER ONE	Waters	1
CHAPTER TWO	Milks	33
CHAPTER THREE	Hot drinks	81
CHAPTER FOUR	Cold drinks (non-alcoholic)	139
CHAPTER FIVE	Alcoholic drinks	188
	Digestif	229
	Bibliography	239
	Index	297

Aperitif

....

Welcome to the wonderful world of drinks. Shots, nightcaps, thirst-quenchers, pick-me-ups, bevvies, brews, nectars, chasers, snifters, grog, libations, Dutch courage, moo juice, Adam's ale and one for the road – whatever your tipple, you'll find it in here. But hang on: there are loads of books detailing individual drinks so what's the point of *Drinkology*? Well, for a start it brings together the whole gamut of the most popular beverages in one place, which is pretty handy, but there's actually a second important purpose.

When you choose a drink do you give it much thought? Many of us think carefully about the foods we eat and how they may affect our health but probably don't give the same attention to what we drink. Think about it . . . do you *really* know what you are drinking? You do? Are you sure? We all consume several drinks every day – perhaps we're just thirsty, perhaps we need something to wake us up, perhaps we need something to relax us at the end of the day. But have you ever stopped to wonder just what those added electrolytes in your bottled water are supposed to do, or why that glass of wine you're sipping might contain shellfish extracts? This is where *Drinkology* really comes in. Every day we're exposed to a stream of inconsistent, confusing and often misleading information about what we consume. This book distils scientific evidence to get to the essence of how drinks are made, what exactly is in them and how they affect the body.

Whether it's a simple glass of water or early morning espresso, the finest champagne or energy drink the morning after, all drinks have an impact on our body in one way or another.[1] According to

---

1 Even if it's just simply refreshment.

Douglas Adams' *The Hitchhiker's Guide to the Galaxy*, the best drink in the universe is the Pan Galactic Gargle Blaster. Its effects are like 'having your brains smashed out by a slice of lemon, wrapped round a large gold brick'. Back on planet Earth nothing quite compares but while there are supposedly superfoods, is there such a thing as a 'superdrink'? We'll see.

Whether you want to discover the true benefits of fermented drinks, find out if sulphites in wine really cause headaches, or are just sick of the pseudoscience behind the marketing of what we consume, *Drinkology* is for you. I've put together a scientific digest of many of the world's most popular drinks for you to savour over time. As well as everyday drinks that we're all familiar with, I take a look at the science behind some of the fads and fashions in the drinks world. Instagram, Twitter, blogs and magazines are filled with attractive images and persuasive articles promoting various concoctions with the promise of health-giving properties. Celebrities line up to endorse them and persuade thousands of their followers to try them, using a jumble of scientific jargon. But are these beverages as good as their claims? (Hint: Gwyneth, Kim and other influencers might like to look away now.)

But *Drinkology* isn't all about the science. This guide is also peppered with historical snippets, world records, fascinating inventions and other random titbits that you can use to impress your friends or win the pub quiz. Think of these added extras as the cherry in the cocktail, or indeed the little umbrella and a sparkler if, like me, you're not so keen on a glacé cherry.

Before we get started, let's quickly down the basics behind drinks.

What are drinks and what physically happens when we consume them?

A drink is simply defined as a liquid, taken in through the mouth, that is consumed for refreshment or nourishment.

When you drink, the liquid passes down the oesophagus, into the stomach then the intestines, where most of the water in the drink is absorbed. Some nutrients in the drink, like proteins or carbohydrates, need to be broken down into a form that can be absorbed into the bloodstream via the small intestine. With alcoholic drinks, a significant proportion of the alcohol is absorbed into the bloodstream via the stomach lining (more on alcohol metabolism in the Alcoholic drinks section). The blood travels to the liver and, there, nutrients are processed and stored, and toxins may be broken down. As required, the remaining minerals and nutrients from your drink, held within the blood vessels, then pass around the body being metabolised and taking effect. Eventually, what's left finds its way to the kidneys where waste products are extracted and converted into urine for excretion.

How much should we drink?
As you know, water is essential for life and is involved in most bodily functions. Our bodies are composed of around 60 per cent water so, for example, a person who weighs 70kg will contain around 42 litres of total body water. We lose large amounts of water every day, so need to replace it to ensure we don't become dehydrated. If we don't take in enough water, there becomes a fluid imbalance in our cells and, given water is essential for maintaining a healthy and functioning body, dehydration can result in a host of symptoms. Severe dehydration can, of course, be fatal.

While drinking too little can cause dehydration, you can also drink too much, leading to something called water intoxication or hyponatraemia. Taking in more water than your kidneys can process also causes a fluid imbalance in the body, which can result in the dilution of salt, or sodium (Na), in the blood (hence 'hypo' [meaning less than normal], 'natr' [referring to the Latin for sodium 'natrium'] and 'aemia' [referring to the blood]). A key role of sodium is to balance fluid in and around your cells, so too much water can lead to extra fluid moving into cells, resulting in them swelling. This can cause health complications such as swelling in the brain, and there have been rare cases of people dying from drinking too much water.

You don't need to worry too much, however, as it is pretty unusual to suffer from hyponatraemia. Endurance athletes are more likely to be at risk as balancing their fluids under extreme physical conditions is more challenging than for the rest of us.

So just how much should we drink each day? Annoyingly, the most accurate answer to this is 'it depends'. Let me explain. Sedentary adults use around 2 to 3 litres of water per day. Our bodies produce a small amount of water (around 250–350ml per day) as a by-product of processing particular nutrients to release energy, but the rest is gained from what we consume. It is estimated that, on average, around 20 to 30 per cent of the water we consume comes from food, with 70 to 80 per cent contributed from our drinks. Every day, we lose water mainly via urine (around 1 to 2 litres), faeces (around 200ml), breathing (250–350ml) and perspiration (around 450ml in a temperate climate). Water losses through our skin and lungs depend upon the climate, air temperature and humidity. In addition, a range of other factors will determine our water needs, such as our age, weight, gender and level of physical activity. Illness can also significantly affect how much we need to drink.

In healthy adults, the feeling of thirst is the primary way of knowing when to drink more. So, for this group of people, thirst is a good guide as to how much to drink each day. This feedback mechanism doesn't work quite so well in children and the elderly and they may not recognise the signs of thirst so readily, so care should be taken to ensure these groups drink sufficient amounts. But what counts towards your fluid top-up? Does it have to be water? Do certain drinks not count? While water is particularly good, as it doesn't contain less nutritious elements, like sugar or alcohol, pretty much all drinks count towards your fluid intake. The main exception is alcoholic drinks. Although they contain water, they increase the amount of water you lose as you urinate more. However, the more dilute alcoholic drinks, such as low strength beer, although not ideal for fluid replacement, can still lead to a net gain of water overall.

Correlation does not imply causation

To get you up to speed, I've tried to include as much information as possible (and possibly too much at times), but I couldn't hope to cover every little detail as there is just too much to say. But before we move on, I just want to make an important point about a lot of the evidence presented in this book.

Much of the scientific evidence referred to is based on observations regarding a relationship, or correlation, between particular drinks and health effects. This means that studies have found a connection between consuming (or not consuming) a certain amount of a drink and a health outcome. For example, people who regularly consume specific beverage X have been found to have a lower risk of particular health problem Y. However, finding a link does not imply that there is a direct cause and effect mechanism going on between X and Y, and there may be alternative explanations. For instance, did you know that there is a correlation between the sales of sunglasses and ice cream? As the sales trend of one increases, so does the other. You can probably guess that buying sunglasses does not cause a person to buy ice cream and vice versa; i.e. while they are correlated, there is no cause and effect. Both factors in this instance are likely linked to the weather, which could be interpreted as a cause of both sales trends.

If you want to see what I mean about this whole issue, there's a wonderful website (and now book) called *Spurious Correlations* by military intelligence analyst and Harvard Law student Tyler Vigen, that takes a light-hearted approach to the matter. He demonstrates beautifully how you can find correlations between the most ridiculous things if you have the right data – it doesn't mean there's a meaningful connection. For example, did you know that the divorce rate in Maine correlates perfectly with the per capita consumption of margarine in the US? Or that the number of people who drowned by falling into a pool correlates with the number of films Nicholas Cage has appeared in?

So, in summary, we need to be careful when we interpret the science behind what we drink. There can be many links between drinks (or even specific compounds within them) and health but whether they are true associations, and what they may mean, are different matters entirely. To confirm true benefits, or indeed harms, of beverages, robust trials (e.g. independent, objective, long term, high quality and large) are needed that compare different consumption patterns, within different populations and investigating specific physiological changes. More research is still required to really drill down into the details and try and unpick whether certain drinks are directly causing health effects.

Enough of the caveats: let's begin our tour of the world's most popular drinks.

CHAPTER ONE

Waters

························

There's no better place to start than water, the foundation of all drinks. Despite being the most natural thing we consume and essential to human existence, the concept of water as a drink is now strangely fashionable. A few years ago US Snowboarders Austin Smith and Bryan Fox launched a company called *Drink Water* to promote the consumption of water. Their idea came in response to energy drinks companies increasingly sponsoring snowboarders and their desire to counteract the marketing of such energy drinks, which they don't believe in. Like the *modus operandi* of many grass roots activists, Smith and Fox first started by writing 'drink water' on their snowboards and making stickers with the same slogan. This eventually expanded into a business selling a range of snowboarding products displaying their mantra. You would think that they also sell bottled water but actually it's just the opposite. As they say: *'We're actually committed to never selling water. We prefer it from the tap.'*[1] They just want people to drink it. Their slogan couldn't be more perfect.

Since we've had a choice in what we drink, water has never been more popular. While a couple of decades ago, the cold drink of choice might have been a can of fizzy pop, we've been increasingly turning back to water. In fact, continuing a long trend of vigorous growth, in 2016 bottled water sales by volume surpassed carbonated soft drinks sales for the first time. And beyond bottled waters, there's ever greater interest in plain old tap water. As environmental concern over the impact of plastic bottles increases, more and more people

---

1 Drink Water, 'Drink water is an idea.' https://www.wedrinkwater.com/pages/reason

are purchasing reusable water bottles to carry around their own tap water. With water now being regularly glugged, and water bottles always to hand when we're on the go, we've probably never been so hydrated!

Obviously, there is water in all drinks so the following section could apply to all drinks (and, in a way, I could just write this entire book about water), but here we deal with drinks that are pretty much just water (with a few exceptions towards the end). Drinks that are really about other contents are dealt with elsewhere. So, let's move on and think about what we really know about water.

Water: in essence

As you know, water is a clear, colourless, odourless substance that we all need to consume to stay alive. It is made up of billions of molecules, each of which is made up of two hydrogen ('H') atoms and one oxygen ('O') atom held together by strong bonds. As such, it is known as H_2O. Despite it covering around 70 per cent of the earth's surface, we can't just drink any water. It needs to be fit for human consumption to avoid ill-health effects. Water that is safe to drink is referred to as drinking water, or potable water, and some water that starts off non-potable is made into drinking water via various treatments and filtering methods.

How is tap water made?

Tap water simply refers to water that comes from a piped source and out of a tap. Here, we'll just concern ourselves with drinking water supplied by our taps.

Locating appropriate sources for our tap water is key. The water we drink is typically derived from lakes, rivers, reservoirs (artificial lakes), aquifers (underground rock layers saturated with water) and boreholes (drilled to reach underground water sources). The water source needs to be fresh, free from toxic chemicals and reliable. Once you have your source, you next need to treat the water and the treatment will depend on the type and quality of the water source.

A number of barriers and disinfection techniques are used to ensure our tap water is safe. This can involve passing it through a series of filters to remove unwanted objects, sediments and small particles; using ozone, carbon and ion exchange to remove micro-organisms, pesticides and metals; and adding a small amount of chlorine to kill off remaining bacteria and organisms, and keep the water safe as it passes through the water system to our taps. Phosphate might be added to reduce the introduction of lead into the water from old pipes, and the water may be passed through ultra-violet light for additional disinfection.

The pH[2] of the water may also be chemically adjusted to reduce corrosion in the system, making the water more stable. Soft water is usually more acidic (i.e. has a lower pH) than hard water, and there-fore potentially corrosive. If the pH is too low, it could cause metals, such as lead, iron, copper and zinc to leach into the water from plumbing fixtures and piping in the water system, as well as leading to an unpleasant metallic or bitter taste. So how do they alter pH? Methods include: adding sodium hydroxide, calcium hydroxide or sodium carbonate; passing the water through an alkaline bed; or removing excess carbon dioxide. (We'll talk more about water pH later – see *Alkaline water*.)

As much as that might all sound rather unappealing, the alterna-tive is much worse. Water contamination is a huge problem in many countries and linked to the spread of preventable diseases like chol-era, diarrhoea, dysentery, typhoid and polio. Globally, contaminated drinking water is estimated to cause over 500,000 deaths from diar-rhoea each year, demonstrating just how important treating water is. Once you start looking into tap water quality, you come across long

2 A scale that defines how acidic or alkaline a solution is; pH 7 is neutral, lower than 7 is more acidic, and higher is more alkaline.

lists of contaminants that have the potential to infiltrate our drinking water. These include: a) physical contaminants, such as sediment or organic particles from the original water source; b) chemical contaminants, such as nitrogen, bleach, salts, pesticides and metals; c) biological contaminants, such as bacteria, viruses and parasites; and d) radiological contaminants, such as caesium, plutonium and uranium. In most of our water, the greatest concern is typically contamination from sewage or animal faeces that may introduce disease-causing organisms, such as *E. coli*. But before alarm bells start ringing, let me stress that only a few of these contaminants might be present in our water at all, and even then in tiny amounts, and water treatment plants effectively deal with many of these potential contaminants before the water gets into the public supply. That's the whole point.

Why does tap water differ depending on where you live?
On any typical day, my hair flows wildly about me in a frizzy mess and my sandpaper-like skin requires inch-thick creams to avoid total collapse, but transport me to a different part of the world for a week and the transformation is bordering on miraculous (think smooth hair and peachy skin, almost). When travelling, we've all had those moments when we take a sip of water and think, 'eurgh, I prefer my home tap water', or realise our hair and skin feel much softer than usual. The reason is often that the water we're exposed to is quite different from where we live and what we're used to.

What makes hard water hard and soft water soft?
Rainwater is naturally soft. Once fallen, it seeps into the ground and picks up minerals from the rocks (particularly chalk and limestone), sand and soil it filters through, which dissolve into the water. The difference between hard and soft water relates to this mineral content. Hard water contains a significant quantity of the dissolved minerals, primarily calcium and magnesium. The higher the level of

these dissolved minerals, the harder the water. This is why the geology of different regions determines the hardness of the water. Although there are varying degrees of water hardness, there is no standard classification of hardness and it varies from country to country. In the UK, for instance, water supplies are typically hardest in the south and east and become softer moving north and west. As you might have guessed, surface water, like that from rivers and lakes, is typically soft and has low concentrations of minerals. So if your drinking water comes from a reservoir, for instance, it is likely to be soft water and if it is sourced from underground, it will be hard.

Can you alter water hardness?
It's the hard water where I live that causes my hair and skin dilemmas, and the problems don't stop there. Those of you who also have hard water will probably understand the never-ending irritation of a furry kettle, limescale spots on chrome taps, smeary windows and the fortune spent on washing detergent. Like many people, you may choose to use a water softener. Rather than improving the quality of your drinking water, water softeners instead aim to prevent limescale deposits caused by hard water and improve water's interaction with detergents, helping them to remove dirt and oils more effectively (soft water requires less soap to produce a lather). Water softeners use the process of ion-exchange to remove calcium and magnesium minerals responsible for causing limescale, and replace them with sodium ions. While calcium and magnesium ions interfere with the way soaps and detergents work, sodium does not. The softer water also helps give skin a softer feel when you wash.

How does ion-exchange work? Water softeners contain porous resin beads that bind to positive ions in the water. The beads start off coated in sodium ions but, as water flows over them, calcium and magnesium positive ions that occur in hard water are attracted to the beads, replacing and releasing the sodium ions. In other words, ions on the beads are exchanged, and the hard minerals are gradually drawn out of the water and trapped in the beads.

Other than cosmetic effects, does the hardness or softness of water have any effect on our health? Many studies have investigated the potential health effects of hard versus soft drinking water. Hard water may offer particular benefits to our health. Although water is not a main dietary source of calcium and magnesium, hard water may be a useful contributor for some people who struggle to get enough from other sources. In addition, there is an indication that hard water may reduce your risk of dying from cardiovascular disease, although the evidence is not entirely consistent. While other health benefits of hard water have been mooted, the data is not compelling. A common misconception is that hard water may cause kidney stones due to its high mineral content, but there is no hard data proving a clear link between the two.

What about soft water? Quite a number of studies have investigated the impact of drinking soft water on health outcomes, due to the relative lack of minerals known to be important for our health. Many such studies found an increased risk of heart problems, but the nature of the research has not been particularly robust and other studies have shown contradictory results. In short, it is still unclear whether soft water may affect your heart. Drinking artificially softened water is not considered unsafe as such but the taste may be affected and be less pleasant than hard water. Having said this, it is usually recommended that if you install a water softener, you retain a source of unsoftened water for drinking and cooking purposes. This is to avoid additional sodium in the diet, which could be a problem for vulnerable groups (e.g. babies or people on a low sodium diet).

What about household water filter jugs? Are they any good?
So you've got your water out of the tap but you're still not quite happy with it. Many of us now have a water filter jug in our home for refining the quality of our tap water. They certainly offer a much cheaper option than stocking up on bottled water and help us do our bit for the environment by avoiding the use of plastic, disposable bottles. But what exactly do home water filters do, and are they really making the water better for us?

The idea of home water filters is to remove unwanted impurities and improve taste whilst retaining beneficial minerals and trace elements. For example, many claim to remove chlorine, traces of metals, bacteria and pesticides, and reduce limescale. To do this, most home water filters contain both a physical, sieve-like filter as well as activated carbon. This is a type of carbon that has been treated with oxygen to make it very porous, enabling it to act a bit like a sponge. These microscopic pores create a huge inner surface area available for the adsorption process, whereby contaminant particles stick to the carbon's surface. When I say huge, I mean really enormous. Just one gram of activated carbon can have a surface area of anywhere up to around 3000m², which means a teaspoon of it can potentially have a surface area larger than a football pitch. With regular use over time, the filters eventually clog up with the impurities and need replacing. Some filters also contain silver, which kills bacteria and can help break down some pesticides.

So, what do they remove? Jug filters are particularly good at removing chlorine and its by-products (e.g. trihalomethanes [THMs] that some are concerned are possibly linked to adverse health effects). These filters are not so good at dealing with many other contaminants, such as heavy metals, fluoride or bacteria. While some claim to remove other potential contaminants, like pesticides, water companies actually have to comply with strict guidelines to ensure such contaminants are removed from tap water during the treatment process. By the time it comes out of your tap there shouldn't be any need for a water filter jug to do this job.

You may be thinking 'they sound great, I'm off to get one' but hold that thought for just one moment. An assessment of water filter jugs by the UK's Department for Environment, Food and Rural Affairs (DEFRA) concluded that under certain circumstances they could lead to a deterioration in water quality, and ANSES (the French agency for food, environmental and occupational safety) agrees there's a potential for problems. Although there isn't sufficient data demonstrating a risk to consumer health, water jug filtration can lead to a decrease in pH, the release of contaminants and a deterioration in the microbiological quality of water (i.e. a build-up of micro-organisms in the jug). To avoid problems, agencies like DEFRA and ANSES recommend that users regularly clean the jug and replace the filter, keep the jug and its water in the refrigerator and consume the water within 24 hours of being filtered.

Removing unwanted tastes and odours is actually the most typical reason for buying a water filter jug but you can usually achieve the same effect by chilling a covered, ordinary jug of tap water in the fridge for a short period. Any chlorine quickly dissipates and the cool temperature dulls our sensitivity to other tastes and odours. In fact, studies have found that most people can't tell the difference between chilled tap water and bottled mineral water. Just remember to throw away any unused water after 24 hours and clean your jug regularly.

Before we move on, it's worth pointing out that other water filter jugs exist that work by methods other than activated carbon and claim to remove even more contaminants. However, there is no single water treatment or technique that can remove all contaminants. The main question is, how pure is your tap water by the time it reaches you and do you really need a water filter? This very much depends on where you live and your expectations about what kind of water you're hoping to consume.

What about fluoridation?

There is something else in water that many people are concerned about: fluoride. Most of us don't give it much thought but for the ones who do (mostly dentists, the health-obsessed and conspiracy theorists), it is a hotly debated issue. Type *fluoridation* into a web search and you'll soon see what I mean.

Fluoride is a mineral that occurs naturally in low levels in drinking water but the amount depends on the rock present where you live. It is also added to some water supplies where natural levels are low, but why? Research has found that the fluoride plays a pivotal role in the prevention of tooth decay. Many areas of the world, including parts of the UK, US, Australia, Canada, South Africa and Singapore, deliberately add fluoride to their public drinking water to ensure a constant level as a measure to improve dental health. Fluoridation programmes have been implemented since the 1940s, following research in the US into the effect of fluoride on teeth.

In 1945, Grand Rapids, Michigan, became the first city in the world to fluoridate its water supply as part of a research study to test whether this would help fight tooth decay. Decades of observational studies up to this point indicated that the different fluoride levels of public water supplies were having an effect on children's teeth. In areas where too much was present in the water, children had brown, mottled or flecked enamel (a condition called fluorosis) but their teeth were surprisingly resistant to decay. Dr H. Trendley Dean, from the National Institute of Health, worked to determine how high fluoride levels could be in drinking water before fluorosis occurred. It was discovered that fluoride levels of up to around one part per million in drinking water did not cause enamel fluorosis in most people and only mild enamel fluorosis in a small percentage of people. Dean then put this information together with the fact that fluorosis was associated with protection against dental caries. He wondered whether in areas with lower fluoride levels in the water you could add fluoride to improve the dental health of the population. This hypothesis needed testing and the good (and trusting) people of Grand Rapids agreed to add fluoride to the public water supply to find out the answer. They monitored tooth decay in around thirty thousand schoolchildren and after eleven years found that the number with dental caries (tooth decay) was more than 60 per cent lower in children born after the fluoridation of the water supply. It was this very significant result that went on to revolutionise dental care around the world.

Since the early days of fluoridation there has been much research to monitor its effects. Numerous studies have consistently found that those living in areas with fluoridated water experience fewer dental caries than those without a fluoridated water supply. Fluoridation of public water supplies is considered a safe and effective public health measure in many parts of the world. However, fluoridation of drinking water is not without its critics and some get very heated about the whole issue. Ever since its introduction, fluoridation of public water supplies has been controversial. Critics put forward a range of concerns including the belief that fluoride causes health harms, that it is expensive, and that it violates individual rights as it removes choice from the consumer. If fluoride is added to the water supply, people have no choice but to ingest it. Champions of fluoridation argue that it is essential that all are given equal access to fluoride to reduce tooth decay, which typically affects socially deprived people the most.

Dental fluorosis is the most frequently cited negative side-effect of fluoridated water. It is true that there are slightly higher rates of fluorosis in areas with fluoride added to the water supply, however the effects are usually mild. While other concerns have been raised about the effect of fluoridation on our health, such as an increased risk of cancer, hip fractures and kidney stones, and brain development, cardiac or metabolic problems, no good evidence supporting these concerns has been found. As one review points out: 'While harms have been proposed, most have no biological plausibility or insufficient evidence to draw conclusions.'[3]

I should point out, however, that contemporary evidence for the fluoridation of water is not that strong. Much of the research into fluoridation schemes took place prior to the 1980s and it may be that changes to our lifestyles and ready access to other sources of fluoride nowadays (such as toothpastes, and fluoridated milk and salt in some countries) could reduce the need for mass fluoridation. The most consistent results find that the strongest benefit is in children and their developing dentition, particularly in deprived areas.

While you might have started now to worry about the added fluoride in your water and whether water companies are infringing your right to choice, your water might not actually be affected. In the UK, for instance, only around 10 per cent of the population are supplied with artificially fluoridated water. There are only a handful of countries where more than half the population receive fluoridated water, including the US, Australia, Singapore, Ireland, Australia, New Zealand and Chile. Most places in the world do not fluoridate public water supplies (e.g. only around 2 per cent of Europeans receive artificially fluoridated water) and some parts of the world that previously did practise fluoridation no longer do so, for a variety of reasons.

What is the difference between bottled waters?
Are you one of those people who feels thirsty at the mere thought of not having any water to hand? Yes? Me too. This constant urge for rehydration makes us a marketer's dream and is probably why bottles of it are now available everywhere we go. Across the world, we drink a vast amount of bottled water and our consumption has

3 Community Preventive Services Task Force, 'Oral health: Preventing dental caries, community water fluoridation.' https://www.thecommunityguide.org/sites/default/files/assets/Oral-Health-Caries-Community-Water-Fluoridation_3.pdf

been increasing. It was estimated that we drank over 370 billion litres of the stuff in 2017.

Who glugs the most?

The country that consumes the most bottled water per person is Mexico, at 67.2 gallons in 2016. By comparison, consumption in the US was 39.3 gallons for the same year.

There are many types of bottled water. They are all drinking water but their labels classify them as different entities. You may not have taken much notice but it can be confusing. So what are the differences between them all? Is there any real difference? Does it even matter? Let's clear things up.[4]

Mineral water versus spring water
Both mineral water and spring water are naturally sourced waters. To be classified as such, they must come from a natural, pure, underground source that is protected from all risk of pollution. The water must be bottled at source, be free from contamination (including harmful micro-organisms) and be safe to drink without treatment. So far, so the same. However, the main difference between mineral water and spring water is that mineral water must have a constant and distinct mineral composition and originate from an officially recognised spring. The regulations for spring water are not quite so stringent.

A limited amount of treatment of mineral water and spring water is allowed so long as it doesn't affect the water's physical and chemical characteristics, safety and microbiological purity. Permitted treatments include: filtering or decanting to remove unstable elements

---

4 In different countries there are different classifications and definitions of the various waters, so here I refer to the UK or EU definitions.

(such as iron, manganese and sulphur that may cause sediments to form during storage); ozone-enriched air treatment to also assist in the removal of unstable elements; fluoride removal in certain circumstances; and the addition or removal of carbon dioxide to create sparkling or still water. In the UK, some additional treatment of spring water is allowed. Disinfection treatments, such as ultraviolet treatment to deactivate bacteria in the water, are allowed in some circumstances but only where they do not cause the water to become unfit for human consumption, by leaving harmful residues for instance.

The vast majority of bottled waters sold in Europe are either mineral or spring waters. Each has its own distinct taste and mineral composition, as determined by the geological conditions of its natural source. The main minerals contained in naturally sourced water include: calcium, magnesium, potassium, chloride, sodium and sulphate.

Water is nothing new

You may think that our passion for bottled water is a relatively recent trend but mineral and spring water producers have been promoting natural waters across the world since the 1950s.

Purified water versus distilled water
Purified water is sometimes also called table water or bottled drinking water. This water is simply drinking water that has been bottled. The water has been purified (e.g. filtered or distilled) to ensure it is safe to drink, which means any water impurities were either removed or reduced to very low levels. It can come from a variety of sources, including out of a tap from public water supplies. An estimated 25 per cent of bottled water in the US is just treated tap water.

Distilled water is created via the process of distillation. In this process, the H_2O is boiled away from any contaminants, such as unwanted minerals and metals. This works because many contaminants have much higher boiling points than H_2O so, as the water is boiled, pure H_2O steam is created and captured. Upon cooling it becomes distilled water and anything that didn't boil off in the steam, i.e. most contaminants, are left behind. A few contaminants remain, such as volatile organic compounds, due to their lower boiling point than water, meaning they also evaporate along with the steam. One of the methods used to purify water is distillation so, in effect, distilled water is also purified water.

Artesian water

Artesian water has a rather exotic sound to it and, indeed, some producers of such bottled water capitalise on this appeal. However, its definition is a little more mundane than you might be led to believe. Artesian water, also known as artesian well water, merely refers to water that comes from a well in a confined aquifer. The water in the aquifer must be under enough pressure to force it up the well to stand at a level above the top of the aquifer. So, artesian water is no different from other groundwater but just comes to the surface in a different way. It has no additional special features or health benefits.

Raw water

Picture the scene: a hipster in a serene yoga pose on top of a rock surrounded by a clear, natural spring, twinkling in the sunlight. He wants you to drink the water. 'That looks refreshing and must be very good for health' you think. The slogan 'Perfected by nature'[5] taps into your now zen-like mood and urges you to get back to basics. Yup, you guessed it, that's how raw water, the latest water craze, is being marketed to those looking for yet another way to live forever.

5 Live Spring Water. https://livespringwater.com

The people of Silicon Valley, California, and elsewhere in the US, have developed a hankering for untreated, unfiltered bottled water, also known as raw water or live water. This in part stems from a fear of consuming artificial chemicals in the public water supply. One company selling this product feeds on this fear by raising concerns that tap water is filled with prescription drugs as well as chloramine and fluoride, which they assert is harmful to health (controversially, the founder of one company considers fluoride to be a 'mind control drug' of no benefit to teeth). Supporters of raw water also believe that, by being untreated and unfiltered, it retains natural probiotics and essential minerals that are otherwise lost during standard treatment processes and so is more beneficial to health. Others are not so convinced of the benefits of raw water, pointing towards its potential to harm, and accuse producers of scaremongering. So what is the evidence behind raw water?

Firstly, producers claim raw water contains probiotics that are essential for our optimal health. There are very specific criteria for what makes a particular bacterium a viable probiotic and the number of bacterial strains that make the grade is low. Without knowing exactly what bacteria are present in the raw water, any claims about its probiotic content are unsupported. Where bacterial strains have been identified for raw water sources with claims about their probiotic nature, it turns out that that there is no hard evidence demonstrating an actual health benefit.

Secondly, the high mineral content of raw water is used as a selling point. Producers tend to give the impression that, by being unfiltered, their water contains natural mineral levels that provide an optimal balance for our health. However, here we do need to point out that not all minerals are good for us. Yes, it may be great to hear that our water contains calcium, magnesium and potassium, for example, but minerals such as lead, arsenic and aluminium may also be present. In addition, we typically get all the minerals we need from our diet, not our water, and the amounts found in water are very small. So, this isn't really a big selling point either.

Thirdly, let's now move away from health claims and consider potential harms of drinking raw water. This is the biggie. Is raw water safe? By rejecting conventional water safety treatments, raw water may contain harmful bacteria (e.g. *E. coli*, *Salmonella*), viruses and parasites (e.g. *Giardia*), as well as other contaminants like arsenic, radon or pesticides, that make people ill. This is what water treatment plants aim to prevent getting into our water supplies. After all, we have to remember that everyone in the past drank raw water, and a huge number of people got sick or died from contaminated water, until the Victorians brought in water sanitation for the masses. There is currently no regulation surrounding raw water so you can't be sure of what you might be getting.

To be fair, not all raw water providers make spurious health claims. Some focus on the purity of their water from contaminants and have been approved by the US Food and Drug Administration (FDA) in leaving their water untreated as they meet stringent safety standards. There may well be some very pure water sources that offer a safe alternative to treated water. If the water source is identified, well protected, stable, has a long-standing safety record, is frequently and thoroughly tested for any presence of harmful contaminants and bottled hygienically at source, then perhaps there is little difference from other mineral waters. While long-term health benefits of such pure water sources have yet to be proved, if anything such sources may at least provide peace of mind to consumers concerned about water treatment processes. The bottom line, however, is probably not to fill up your water bottle from a clear spring just because it looks fresh and inviting; ultimately, you don't know what (or who) may be lurking in the water (who knows where that hipster has been?!).

Alkaline water

You may have come across something called alkaline water, sometimes also referred to as ionised alkaline water. It is often touted as being good for our health. Most commonly, you can buy it ready bottled or purchase a device that purports to transform regular water into alkaline water, via ionisation. I was recently fascinated by the many gadgets available for ionising water when travelling on a long-haul flight via Singapore airport. The in-flight magazine had a couple of products on offer and more could be purchased at the airport. I love a gadget and am often tempted by something novel when I'm bored out of my mind on a flight. What is this wizardry, I wondered? Does it really work? Cue the Google search.

As you might expect, alkaline water has a higher pH than regular drinking water. Some people believe it can neutralise the acid in our bloodstream and, by doing this, it can help prevent diseases like cancer and heart disease as well as provide benefits for our bone health. Advocates also claim that alkaline water has anti-ageing properties, hydrates us more effectively, supports our immune system and can even aid weight loss. Sounds amazing! (Out of curiosity, I tested the pH of my home tap water and was excited to discover that it is around pH 8–8.5; wonderfully alkaline. With my regular consumption of this water over many years, surely I will live ailment-free, forever?) But, is there hard evidence to support the hype? In summary, no. In 2016, a comprehensive review of studies concluded that the health-promoting claims for alkaline water are not supported by evidence and are, quite frankly, unjustified.

The thing is, unless you have a specific underlying condition such a kidney or lung disease, your body is very good at maintaining a healthy pH by itself. Everything we eat and drink goes straight to our stomach. There it meets the highly acidic stomach juices needed to break everything down and kill off bacteria. Any alkalinity from what we consume disappears, rendering any potential benefits unlikely. You would have to consume a vast amount of alkaline water to neutralise your stomach's acid. The most common acid in our body is actually carbon dioxide (CO_2). When you produce too much of it, your body merely gets rid of it to maintain the pH balance. As blood flows through our lungs, any excess CO_2 is removed and exhaled. Our kidneys also help to regulate any pH imbalances as necessary.

So much for the exciting gadgets. Try reading about the ionisation process and you'll discover a whole load of nonsensical gobble-dygook; more science fiction than science fact. Many of the terms used are not grounded in science but are marketing inventions to bewilder the public into buying the products. Good job I decided not to try one.

What is sparkling water?

Apparently, the average person in the UK drinks 5.9 litres per year of sparkling water (I suspect that in reality it is more likely that a small number of people drink a lot of the stuff and a large number don't touch it.) Sparkling water, also known as carbonated water, is water that contains carbon dioxide (CO_2). It is either naturally carbonated, has its natural carbonation adjusted by the addition or removal of CO_2, or is still water into which CO_2 gas has been dissolved.

The presence of CO_2 means that some sparkling waters have a lower pH, and are therefore slightly more acidic, than still water. This has led to concerns that consuming large amounts of sparkling water may cause erosion of tooth enamel. However, there is no good evidence for this, with the few studies available showing little effect of sparkling water on teeth. Although drinks with a low pH may have a higher erosive potential, you would probably have to drink a lot of sparkling water, regularly and over a long period, for this to be a problem.

Other health concerns, such as our bones being leached of calcium, or stomach problems resulting from the fizz, are also refuted by the scientific evidence. While other fizzy drinks have been linked to poorer bone health, this has not been found in carbonated water. It seems the problem may lie with other constituents of the fizzy drinks in question, not present in sparkling water (more on this in Chapter 4, cold drinks). As for the stomach, some research has actually found there to be benefit in sparkling water over still water. Studies in people with tummy complaints, such as dyspepsia or constipation, found that sparkling water, versus still water, was helpful in relieving symptoms. In fact, many people find sparkling water settling for an upset stomach.

Some people report feeling full or bloated after drinking sparkling water and there's evidence to support this, but whether it can be a handy trick to help lose weight is another matter. While many believe the feeling of fullness caused by sparkling water to be helpful in reducing the amount of food we consume, there's evidence to the contrary. Carbonated water may actually increase appetite. A small amount of research has found that consuming carbonated drinks increases the hunger hormone, ghrelin, leading to an increase in appetite and more calories consumed. However, the increase has mainly been found with sugary carbonated drinks and only a very small increase found with sparkling water. Why carbonated drinks increase appetite is still under investigation but hypotheses include: a) that the drink's CO_2 causes a signal in the top of the stomach to release ghrelin, resulting in hunger; and b) the gas may cause the stomach to bloat and stretch, stimulating cells to release ghrelin.

There isn't hard data supporting the notion of a negative health impact from sparkling water. On the whole, drinking sparkling water recoups the same health benefits as drinking still water. If you're still not convinced, then just drink it in moderation and try using a straw to avoid prolonged contact with your teeth.

What are enhanced waters?
As a species that is constantly reinventing the wheel, our approach to water is no different. We just can't leave well alone and, in the eyes of many, there's always something that can be added to improve a

product. Hence, the evolution of enhanced waters and stretching scientific concepts to the extreme.

Oxygen water

At the risk of annoying some of you, I have to say that in my opinion the concept of oxygen water (or oxygenated water) is possibly the most ridiculous of the enhanced waters. It is exactly what you think it is – water with extra dissolved oxygen added to it. What's the point of that, you may ask? Purveyors of oxygen water claim that the extra oxygen enhances athletic performance, slows the ageing process and increases energy, amongst other benefits. The idea is that the extra oxygen in your system helps your heart and muscles work better, leading to a host of other health improvements including detoxing your system. The target market of this water was originally athletes looking for a competitive edge. This stemmed from the observation that breathing air with a higher than normal oxygen concentration during vigorous exercise slightly improved performance. The market since expanded to those looking for products to enhance their health and wellbeing. But there is quite a big physiological difference between inhaling oxygen and drinking it. While our lungs are designed to process oxygen efficiently, our intestines are not. There's no real evidence to show that oxygen is absorbed from drinking water (unless you have fish gills). So, regardless of whether the benefits of extra oxygen are real, if it doesn't make it into your bloodstream it's all rather irrelevant.

At the end of the day, if you feel like you need a little more oxygen then take some deep breaths. There's far more oxygen in a single breath than you'll gain from a bottle of oxygen water.

Added electrolytes

You may have noticed that some bottled waters proudly proclaim the inclusion of added electrolytes, but just what are they and do we really need this extra addition?

Electrolytes are salts and minerals, such as sodium, potassium and chloride, that conduct electrical impulses in the body. They are important in the control of physiological functions and an even balance needs to be maintained for your body to function properly. While severe dehydration and certain diseases can cause significant electrolyte imbalances, a typical scenario that may affect your electrolyte balance is during intensive exercise, when they're lost through sweat, or when you have a rapid loss of fluid, such as after prolonged or repeated vomiting or diarrhoea. You might grab an oral rehydration solution from the pharmacist to pick yourself up after a bout of illness, but what about when you're not ill? Are added electrolytes in your mineral water of any use to you then?

Products that mention added electrolytes tend to be water that has been distilled to remove its minerals then had minerals specifically added back in. What's the point of that, you may wonder? Well, the idea is that the water is made more 'clean' via the distillation process to which you can then add 'just the right amount of electrolytes'.[6] You might be thinking there is supposed to be some health advantage from this, but even the most well-known brand (*Glaceau Smartwater*) is shy to make such a claim and only goes as far as to state that the electrolytes in the water are there just to improve the taste. In fact, when you look at the mineral (i.e. electrolyte) content of this particular enhanced water, it contains less than most other mineral waters and even regular tap water.

We naturally get the electrolytes we need from our diet and they are not usually a concern for most of us. There's no real evidence that topping up with extra electrolytes from enhanced water will have any benefit. Unless you're ill or intensively exercising, then drinking plenty of tap water, that naturally contains minerals, is probably all that's needed to keep you sufficiently hydrated.

6 Coca-Cola, 'GLACEAU Smartwater.' http://www.coca-cola.co.uk/drinks/glaceau-smartwater/glaceau-smartwater

> **Sweating salt**
>
> Athletes intensively training in extreme conditions, such as hot weather or high altitude, can lose around 3 litres of sweat per hour. As sweat contains sodium, sodium losses for professional athletes can be significant. For example, for every 5 litres of sweat produced, an athlete will lose around 10g of salt as sodium. That's the equivalent of two and a half teaspoons worth.

Vitamin waters

We may not need extra electrolytes but with our busy lives these days maybe we could all do with a few more vitamins in our diet? If so, isn't it great that we can now top up via a few quick swigs of vitamin water?

Vitamin waters refer to various blends containing water, particular vitamins, sugars or sweeteners, colours and flavourings. They may sound like a panacea for health but in reality there are a number of issues. First, it's well known that most of us don't actually need to consume more vitamins as we're usually able to get the vitamins and minerals we need from a balanced diet, and taking too many may even cause some harm. An excess of certain vitamins, such as vitamins A and E, may be linked to a slight increased risk of cancer and heart disease and very high doses of vitamin B6 can lead to nerve damage. Second, even if we did need more, you wouldn't actually consume as many vitamins as you think. Fat-soluble vitamins (A, E, D and K) will only be readily absorbed if they are dissolved in dietary fat so you really need to consume them with food at the same time. Vitamin drinks mainly promote the inclusion of the B vitamins and vitamin C. Water-soluble vitamins, such as these, can degrade when they're in water for a long time, and this degradation can be exacerbated by heat and light. So, depending on when the drink was produced and how it has been stored before it reaches your mouth, who knows how many vitamins you're left with. In all likeliness, most of any vitamins will probably go straight through you and end up as expensive pee! Finally, it's not just vitamins that are added to the water. Even if you absorbed every little vitamin in the bottle, you will also be ingesting a large amount of sugar or sweetener. The most prominent brand contains around 32g of sugar per bottle, the equivalent of eight teaspoons worth. There are now clear links between sugary drinks and harms to our health, which I'll talk more about later (see Chapter 4, Cold drinks).

Many people have objected so strongly to some of the claims made about the main brand, *Vitaminwater* from Coca-Cola, that formal complaints and lawsuits have been taken out against the manufacturer. In 2009, a number of adverts about the product were banned by the UK's Advertising Standards Authority as it was deemed that they made inflated claims about the drink's benefits. In 2016, in the US, the company settled a lawsuit about other misleading health claims it had made about the product.

You've probably guessed by now that waters with added vitamins are generally a waste of money, and may even pose a potential harm to health from the other ingredients they contain.

Rosemary water

We're all very familiar with rosemary – a fragrant herb hailing from the Mediterranean, enlivening the food we eat – but it may offer us more than just a flavour boost to our meals. There is an increasing trend for including rosemary in everything to improve our health, from skin creams, to tea, to toothpaste. And water is no exception. While a number of manufacturers are adding rosemary to their drinks, there's one main company that is making just rosemary water. Its ingredients are spring water and 4 per cent rosemary extract (containing rosmarinic acid, eucolyptol [1,8 Cineol] and glucosamine). More on these in a moment.

Rosemary has been exploited for its medicinal properties for thousands of years and numerous health claims have been made about its powers. The manufacturers of rosemary water were inspired by the Italian coastal village of Acciaroli (I watched the company's promotional videos all about this so you don't have to). It had been observed that nine out of ten locals lived to over a hundred years old and in very good health. Their lifestyles hadn't been found to be especially virtuous, so it led some observers to ponder on what it is that's helping these folk live healthy lives for so long. You can probably guess where this is heading, as scientists visiting the village discovered that the people there consumed a lot of rosemary.

There's some evidence to suggest that rosemary may aid memory, improve our cognitive performance and slow our inevitable cognitive decline as we age. Rosemary has also been found to have anti-inflammatory and antioxidant properties. It may have a role in slowing the spread of cancer cells, may help in the management of diabetes and may promote eye health. The German Commission E, a governmental regulatory agency, has approved the use of rosemary for the treatment of dyspepsia (internal use), rheumatic and circulatory problems (external use). In addition, under European Union (EU) regulations, rosemary is allowed to be listed as an antioxidant by food companies following a review of the evidence. This all seems promising, but what about the benefits of rosemary water itself? A small study published in 2018 assessed the effects of rosemary water in eighty adults randomly allocated to consuming either rosemary water or plain water. After subjects completed a number of computer tasks, it was found that the rosemary water group fared slightly better on the cognitive tests. This novel finding could be the start of something interesting, but the study has several limitations and, to part me from my cash for a bottle of rosemary water, I'll need more to go on than this. At the time of writing, this seems to be the only research demonstrating a benefit of rosemary water, and the long-term effects are still unknown.

Although unlikely to occur when casually imbibing rosemary water, taking large doses of rosemary can have nasty side-effects, such as vomiting, seizures and coma and may cause miscarriage in pregnant women. As not enough is known about its safety during pregnancy, it is advised that pregnant women avoid consuming large quantities of rosemary.

Until large, independent, robust trials investigate the short- and long-term effects of rosemary water, many questions remain. Are there enough active ingredients in the water to make a difference to health or would you have to drink a barrel load every day to gain a long-term benefit? Are the active ingredients of rosemary effectively absorbed and used by the body when prepared in water? If so, how does this all work? Are the active ingredients stable in water or do they degrade before you drink it? Are there any unwanted side-effects of long-term use? Is rosemary water the anti-ageing miracle we're all hoping for or a marketing ruse to part us with our cash? Time will tell.

On the rocks

People in the US love ice in their drinks. If you've been there, or live there, you'll be familiar with cold drinks being topped to the brim with ice. In other countries, attitudes towards ice are perhaps bordering on lukewarm. Back in nineteenth-century North America, where natural ice was plentiful, it was harvested and sold for industrial purposes. One of the pioneers of the ice-harvesting industry eventually had the idea of persuading people to chill their drinks with ice, in order to expand his market. It didn't take long for the masses to develop a preference for chilled drinks that better suited their hot and muggy summers. The harvested ice was also shipped across the Atlantic and became a fashionable, must-have item in Victorian

England, but the craze didn't last long as ice was a luxury that most people couldn't afford. Even when artificial refrigeration came along, the Brits hadn't developed a great desire for ice-cold drinks and things haven't changed much since. There could be many reasons for this, not least because we just don't get the long, hot spells of weather experienced in places like the US, so find very icy drinks a bit, well, too chilly. Nevertheless, people in most countries will likely spy a chunk of ice bobbing in their cold drink at least once in a while so it's worth some discussion in a book about drinks.

I probably don't have to tell you (but will anyway for completeness) that ice is simply water that has been frozen. Most of the ice we consume is made at home by ourselves, in the ice machines of pubs, bars, restaurants and hotels, and in industrial manufacturing plants. Ice, by its nature, looks clear, clean and refreshing but it may not always be that pure. Many of us are aware that we should probably avoid ice in our drinks in some less-developed countries as the water from which it is made might be contaminated. However, we don't necessarily give the same thought to ice made closer to home.

Studies of ice from a range of sources, whether home-made, from ice machines in pubs or from industrial plants, have found microbial life contained within it. Some bacteria, viruses, yeasts and moulds found in ice are known to cause ill health. The thing is, freezing them doesn't kill or inactivate them. As an example, a study of food establishments in Las Vegas found that one third of the ice samples collected exceeded the Environment Protection Agency's limits on bacteria concentration, and over two thirds contained coliform bacteria, which indicate a possible presence of harmful bacteria. Interestingly, one study of bacteria in ice found that the microbiological risks actually decreased once the ice was put into alcoholic or carbonated beverages, whereas a different study looking at moulds and yeasts found that they retained their viability even when added to alcoholic and soft drinks. When you start to think about it, you realise how easy it is for contamination to occur. Consider the person working behind the bar who fills a glass with ice by dragging it, bare-handed, through a bucket of cubes. And that person who used a scoop to pick up the ice, also bare-handed, then drops the scoop back into the pile of ice when they're finished. In an instant, anything lurking on the person's hand gets transferred to the ice. These days, people tend to be pretty conscientious about hygiene when handling food but not as much care necessarily goes into the handling of ice or drinks in general.

So, ice has the potential to be a consumer hazard and in commercial use could be a vehicle for instigating outbreaks of gastrointestinal diseases. In fact, outbreaks of norovirus, *Salmonella*, hepatitis A and *E. coli* have all been linked with ice consumption. Of course, we need to put this into perspective as the majority of people consuming ice in their drinks when out and about don't fall ill. Still, you do need to have faith that the establishment making the ice carries out effective and regular maintenance, cleaning and disinfecting of their ice machines, and has good hand hygiene when preparing drinks.

Microbiological risks aside, does consuming ice have any other effects on our health? Although not all research findings have been consistent, drinking ice water has been associated with an increase in the heart rate of healthy people, and reduction in gastrointestinal temperature and forehead temperature. Such temperature changes may be why ice-cold drinks offer a refreshing sensation. Some believe that adding ice to their water will help them lose weight as the body has to burn extra calories to keep warm. However, this has been shown to have only a very minimal impact on our metabolism and so doesn't point towards an effective weight-loss strategy.

Ever heard of 'brain freeze' or 'ice-cream headache'? Maybe you've experienced that pain when you consume too much of, and too quickly, something that is very cold, like ice cream or an icy drink. The pain can range from a pressing to a throbbing or a stabbing headache, and the headaches tend to be located towards the forehead and temples. Studies looking into this phenomenon compared taking icy drinks with just placing ice in the mouth to see if there are different physiological responses. Different sensations were caused when iced water was drunk versus just pressing ice cubes on the palate, with a more intense pain, but of shorter duration, being produced by the iced water. It also seems that people who've previously experienced migraines and other forms of headache are more susceptible to brain freeze from icy drinks. But what is actually causing the headache? It's thought that it occurs when something very cold comes into contact with our mouth's palate or back of the throat, causing small blood vessels to constrict then rapidly dilate. Pain receptors near the vessels detect discomfort and send a message to the brain via the large trigeminal nerve. Instead of feeling the pain in your mouth or throat, it is referred (i.e. felt elsewhere in the body) as the brain thinks the pain is coming from the head.

One final point on ice: if you are going to have it, how much to put in your drink? Some people don't like to have too many ice cubes in their drink in the belief that this will dilute it. However, physics tells us that the more you put in, the slower they will be to melt and so it would take longer for your drink to be diluted than if you pop just one or two cubes in your glass.

A quick word on the bottles themselves
Should we worry about harm from the plastic bottles containing our water? In recent years, increasing concerns about a plastic used in bottles and other food containers, called bisphenol A (or BPA), have been raised. BPA may be found in some hard plastic bottles, the sort you might reuse. It has been found that BPA can seep into the food or water it is containing and there are fears about the potential harm to health. While studies have found that BPA can cause kidney and liver damage in rodents, this is at very high doses (over a hundred times the tolerable daily intake set by the European Food Safety Authority (EFSA)). Research is ongoing but EFSA, as well as the US

FDA, have so far concluded that BPA poses no risk to human health as exposure levels are so low. Of course, if you're still worried, you can reduce your exposure by opting for BPA-free packaging (which is often labelled as such).

Having said all this, the bottles of most bottled water that you buy do not contain BPA and are in fact made of another plastic called polyethylene terephthalate (PET). Some think that temperature has an effect on whether chemical components of plastic bottles, such as formaldehyde and trace metals, leach into the water they contain. One research team looked at whether there was an effect in PET-bottled waters and found that warmer temperature, along with the presence of CO_2, increased the release of such chemicals. However, no toxic effects were observed. There have also been head-lines warning of the dangers of leaving plastic water bottles in the sun. Concerns have been raised that this could cause chemicals in the bottles to leach into the water, which could result in changes to our hormones, leading to cancer. The same researchers who looked at the effect of temperature went on to consider the effect of direct sunlight. They found that while some migration of particular chemicals (e.g. formaldehyde) occurred in carbonated water (but not still water), when exposed to up to ten days of natural sunlight, the amount was very small and there were no toxic effects at this level. So, while chemicals from plastic water bottles can transfer into the water, it seems they are at very low levels and unlikely to cause harm. Cancer Research UK agrees that there is no good evidence that using plastic water bottles increases the risk of cancer.

Manufacturers state that you should not reuse plastic bottles after consuming the water. The main reason for this is not necessarily because of concerns about chemical contamination as the bottle suffers wear and tear but mainly about the build-up of potentially harmful bacteria. Let's be honest: how thoroughly do you really wash out that water bottle every time you use it? Maybe a quick rinse then fill it up again? The water looks pretty clean, right? Wrong.

Bacteria enter water bottles very easily once they're opened and the bottles, with their ridges, tiny cracks and lids, provide a good environment for them to grow and thrive. Biofilms (i.e. a slimy layer of bacteria and other micro-organisms) can form quickly inside and need a good scrub to remove. (If you have a water bottle that you tend to just rinse out under a tap with no soap, put your finger in and touch the inside – does it feel a bit slimy? That could be a biofilm.) These micro-organisms are probably more likely to pose an immediate harm than any tiny amounts of plastic chemicals in the water.

Is there anything in water we should worry about?
Actually, yes. The pharmaceuticals in our drinking water may be cause for concern. Water supplies around the world contain traces of drugs, whether legal or illegal, from antibiotics, to hormones, to antidepressants, to painkillers, to mind-altering drugs . . . the list goes on. Drugs get into our water supply via a number of routes, from our excretion of them, from uncontrolled drug disposal (i.e. washing them down the toilet or putting them in with our rubbish), from poorly controlled manufacturing processes, and from agricultural sources – any drugs given to animals may pass directly into the ground or wastewater, or pass into the water supply via manure spread onto the land. Some traces of drugs degrade before they reach us and others are removed during routine water treatments, but a proportion may remain. Water companies are not yet able to remove all traces of pharmaceuticals from our water supplies. It is not yet known what effects these may be having on our health. The amounts we're talking about are very tiny and on a day-to-day basis the risk of harm is likely to be small, but over a lifetime effects may build up. There are substantial gaps in our knowledge on the issue and, as such, organisations, like the World Health Organisation, are unable to provide guidance on the issue but, with information to suggest potential harm to health, they are actively monitoring the situation.

Only time will reveal what the consequences may be. So what can we do about it? On an individual level there isn't really anything we can do. We can't avoid it as every water source is affected; even the purest springs. As Professor Tim Spector points out, in his excellent book *The Diet Myth*, even bottled water may not be safe as most varieties tested have been found to contain bacteria that have been exposed to antibiotics, for instance.[7] He's concerned that even the tiny amounts of antibiotics we're regularly consuming in our food and drink could be altering our gut bacteria for the worse, with potentially dramatic consequences. It will take a combination of reduced use of antibiotics in agriculture and human medicine, changes in how we fertilise our crops, better regulation in the disposal of medicines, and advances in water treatment to make a significant difference to the content of our water. For this to happen will take government and industry action, far more research and time.

Beyond the presence of drugs, many bottled waters have also been found to contain tiny particles of plastic, or microplastics. One investigation analysed 250 bottles bought in nine different countries (including well-known brands such as Evian, Dasani and San Pellegrino) and found that nearly all of them contained microplastics in the water. Researchers have also previously found microplastics in tap water and beer, among other sources. There is currently no evidence that ingesting microplastics is harmful to health, but research is ongoing to identify the potential implications. Finding a way to reduce their presence would also be welcome.

In the meantime, for most of us there's no point in worrying too much about these things we can't avoid; the worry itself could be more likely to harm our health!

7 Spector, T. (2015), *The Diet Myth: The Real Science Behind What We Eat* (Weidenfeld & Nicolson)

More money than sense?

The world's most expensive bottled water in 2017 was priced at a whopping US $60,000 for 750ml. *Acqua di Cristallo Tributo a Modigliano* was presented in a bottle made of 24-carat gold based on artwork by the Italian artist Modigliano. The water was described as a blend of natural spring waters from Fiji and France, plus glacier water from Iceland. It also contained 5mg of gold dust.

CHAPTER TWO

Milks

....

In 1998, Ashrita Furman from New York walked for 23 hours and 35 minutes around a track with a milk bottle balanced on his head to land a Guinness World Record, for covering the greatest distance while undertaking this balancing feat; a total of 130.3km (80.96 miles). He's not the only one who holds a milk-related record – there are lots of random records to try and break whether hula hooping with a milk bottle on your head, squirting milk from your eye or donating the most breastmilk. Regardless of what motivates people to do these things,[1] milk holds a central place in our lives.

Milk: in essence

There are a wide range of different milks now available, but what exactly is milk? Milk is typically defined as a white fluid that is rich in nutrients, secreted by the mammary glands of female mammals to feed their young. This points to animal milks as being what is referred to but non-animal milks are increasingly popular. We talk of 'almond milk' or 'soya milk', for example, but given the common definition we've just described are non-animal milks really milk? Such a question was brought to the fore in the summer of 2018 as the FDA suggested that only animal milks should be able to identify as 'milk'. This comes on the back of years of complaints from dairy farmers in the US, UK and elsewhere that non-animal milk producers are capitalising on the milk brand and potentially

---

1 In Ashrita Furman's case the motivation appears to be just to break records for the sake of it. He is the Guinness World Records' most prolific record-breaker and has set more than six hundred records in his time so far (Guinness World Records, 'Ashrita Furman: Guinness World Records' most prolific record breaker.' http://www.guinnessworldrecords.com/records/hall-of-fame/ashrita-furman).

misleading consumers about the nutritional content of their products. Non-animal 'milks' certainly don't fit the traditional definition (as FDA Commissioner Scott Gottlieb commented in 2018, *'An almond doesn't lactate, I will confess.'*[2]). Further to all this, the European Court of Justice ruled, in June 2017, that purely plant-based products cannot be marketed in the EU with designations such as 'milk', which is why you'll typically see dairy alternatives described in terms like 'oat drink', 'milk alternative' or even 'mylk'. (However, I'll be referring to these plant-based products as milk, for convenience.) While these measures help to clarify the distinction between products, perhaps animal farmers are mostly trying to stem the downturn in the popularity of dairy products.

Traditional milk consumption in the US and UK has been declining for some years now. In 2017, the country that consumed the most milk by far was India, at 65.2 million metric tons, and it is predicted that by 2026 it will be the world's largest producer of milk. During the same year, the EU as a whole consumed around 35.5 million metric tons and the US a measly 26.3 million. Of course, India has a very large population which skews the picture slightly. When we look at milk consumption per person, Ireland actually drank the most (125 litres per year, in 2016), closely followed by Finland (120 litres), whereas people in India only consumed 47.4 litres each. While the amount of milk and milk products consumed is higher in developed countries, the gap with many developing countries is narrowing as they increase their consumption of this traditional foodstuff and we reduce ours. The rising demand in other countries is particularly pronounced in East Asia, according to the Food and Agriculture Organisation of the United Nations, especially China, Indonesia and Vietnam.

Today's decline in dairy consumption for many of us is in sharp

2 Irfan, U. '"Fake milk": why the dairy industry is boiling over plant-based milks.' https://www.vox.com/2018/8/31/17760738/almond-milk-dairy-soy-oat-labeling-fda

contrast to much of the twentieth century. As Education Secretary in the early 1970s, former UK Prime Minister Margaret Thatcher abolished free school milk for the over sevens in schools. This led to her being called 'Margaret Thatcher Milk Snatcher' by the Labour opposition and the nickname stuck. The nation was horrified by this policy move, prompting angry protests and a national newspaper (*The Sun*) calling her the most unpopular woman in the country. It was seen as a mean move, taking food out of the mouths of children and potentially risking harm to their health, but it is unlikely that such a reaction would be seen today.

The proposal to introduce free school milk took place in the early 1900s in the UK, with a few schools starting to bring it in, but it wasn't until 1946 that the provision of free school milk was enshrined in law so all schoolchildren could benefit. Change to the scheme actually came prior to Margaret Thatcher's famous intervention as free school milk was abolished in secondary schools in 1968. Provision of free milk in schools waned over subsequent years, due to various government policies. It has been suggested that the original introduction of free school milk was primarily driven by a desire to provide a consistent demand for the stagnant dairy industry of the time, rather than specifically tackle child malnutrition as it was promoted. A recent move to bring back free school milk was mooted in 2016. This time the aim was made clear: to boost children's health as well as support hard-pressed milk producers. It may be that history is repeating itself.

Regardless of motives, it is likely that such a move would be far less popular than in the past. Dairy is no longer seen by so many as the nutritional wonder food it once was, but selling the concept to the masses won't be anything new. Coincidentally, the government also had to sell the benefits of milk to people back at the very beginning of the twentieth century when this whole saga began. In the 1900s, milk had an image problem, being regarded as expensive, potentially infected with disease (it was associated with cases of tuberculosis and other infections) and too variable in quality. Those in charge had to persuade parents and children that milk was a safe and nutritious drink, essential for healthy growth, but their message got a significant boost in the 1920s and 1930s when, concurrently, research into vitamins blossomed. This lent the campaign some scientific backing as milk particularly benefited from this research, as it was discovered that it contained more beneficial nutrients than was previously known. It was increasingly seen as a complete food containing fats, proteins and vitamins; and the rest, as they say, is history.

You can't have failed to notice that supermarket dairy sections have expanded beyond recognition in the last decade. Sitting next to your typical skimmed, semi-skimmed and full-fat cow milk, are bottles and cartons of milks produced from a wide array of plants. Global sales of these milk alternatives more than doubled between 2009 and 2015, to a mighty $21 billion. There has been, and is continuing to be, a huge shift towards plant-based milks. They are no longer the preserve of vegans, the dairy-intolerant and those morally conscious about possible harms of the dairy industry, as a wider audience is also now questioning the benefits of traditional dairy milks versus plant milks.

World Milk Day

There's a special day for everything now and milk is no exception. The first World Milk Day was held in 2001 and it continues to take place on 1 June each year to shine a spotlight on milk and the milk industry.

Baby milks

For all of us, the first drink we take in life is milk. Without it, we simply wouldn't grow and develop. But we don't all get given the same thing. Some of us are raised on breast milk and some on formula milk, some babies even have both, but what is the difference and does it matter which you consume?

What is in breast milk?
Breast milk is billed as the most nutritious food a baby can have. There are different types of breast milk depending on the stage of lactation and there are three main stages. In the first few days after birth colostrum is secreted (a yellowish, sticky milk), which contains more protein, less fat and a number of immunising factors for the newborn. Transitional milk comes next and is produced from about day eight to twenty; this is where the milk transitions to mature milk. Mature milk finally comes in after about twenty days onwards and varies within, and between, individuals in terms of its content. Beyond individual variations, breast milk is not a constant entity but can change over time. Breast milk can change to meet a baby's individual nutrition and fluid needs, and can vary across a day. It even changes during a feed as, for example, the fat content increases as the feed progresses. Mature milk continues to provide immune factors and non-nutritional elements to the baby.

Breast milk comprises everything a new baby needs for healthy development. This includes water, proteins, fats, carbohydrates (mainly lactose), minerals (particularly sodium, potassium, calcium, magnesium and phosphorous), all the vitamins (except vitamin K), although some only in small amounts, and trace elements. As well as essential nutrients, breast milk also contains thousands of distinct bioactive molecules that help protect against infection and inflammation and support the immune system, organ development and promote healthy microbes. These important components include hormones, growth factors, antimicrobial factors and digestive enzymes, amongst others.

Human breast milk contains more sugar (lactose) than cow milk (approximately 6.8 per cent versus 4.9 per cent) and a similar amount of fat (both around 4.5 per cent) but less protein (1.1 per cent versus 3.6 per cent). In fact, when you compare the composition of milk across species, our milk is particularly low in protein. Whale milk, for instance, contains a huge amount of fat (34.8 per cent) and protein (13.6 per cent) but little sugar (1.8 per cent).

Breastfeeding rates around the world vary widely, with the UK having one of the lowest rates in the world. Exclusive breastfeeding is universally recommended for the first six months of life, with most countries also promoting continued breastfeeding along with complementary foods up to two years or beyond. Despite this, a study in 2016 estimated that only 34 per cent of babies in the UK were receiving some breast milk at age six months compared with 49 per cent in the US and 71 per cent in Norway. At one year old, fewer than 0.5 per cent of babies in the UK were receiving some breast milk. In fact, UNICEF produced a report in 2018 highlighting that the world's richest countries typically have the lowest breast-feeding rates. There are multiple and complex reasons behind why women don't breastfeed but does not breastfeeding really matter? Is there hard evidence to show that breastfeeding is superior to feeding your baby infant formula instead? We'll come onto this shortly.

What is formula milk?
The breast milk substitute, formula milk (also known as baby formula or infant formula), is typically made from skimmed cow milk that has been treated to make it more suitable for babies. There is a wide range of formula milks available but in general it comes in two main forms: as a dry powder that you make up with water or as a ready-to-feed liquid formula.[3] It is a big and growing industry. In 2019, it's estimated that the global market value of all baby formula milk is around $70.6 billion. With low breastfeeding rates in many developed countries, formula milk provides a convenient alternative and is the staple diet for millions of young babies.

3 There is also a form of concentrated liquid formula that can be diluted with water.

It is impossible to replicate exactly the make-up of breast milk as it is a complex fluid that is not only variable among individuals but also within an individual over time, but formula milks aim to come as close as possible and are the only recommended alternative. Along with the essential elements of cow milk, regular formula milks contain a range of additional ingredients to try to provide a similar balance of key nutrients that are in breast milk. These added ingredients include: vitamins and minerals, fats (as vegetable oils, e.g. palm oil, coconut oil, sunflower oil, rapeseed oil and/or soy oil, and fish oil), emulsifiers and amino acids. Some also contain probiotics and prebiotics, and advances in technology are enabling the development of other bioactive compounds for inclusion in formula milks. Individual brands of formula milk vary, not just in how they are made but also their levels of protein, fats and micronutrients, so you'll need to look at the ingredients labels for a list of exact contents. Formula milks have to pass strict regulations regarding their contents, and no evidence has been found that one type is better than another for babies, so whichever brand parents choose to go with it should be safe and nutritious.

In the UK, most brands refer to the type of protein they are most dominant in (i.e. whey dominant or casein dominant). Whey-dominant milks are easier to digest so are recommended for younger babies (sometimes called first milks or stage 1). The ratio of whey to casein proteins in these milks is about 60:40, approximately the same ratio as in human milk. Casein-dominant milks are aimed at older, hungrier babies because they take longer to digest (sometimes called second milks, stage 2 or follow-on milks). The ratio of whey to casein in this formula is quite different at 20:80, similar to cow milk.

Various brands and types of follow-on milk are available. This is infant formula that is designed for babies over the age of six months and is intended for use as a liquid part of the weaning diet. There are clear warnings from manufacturers and health bodies that these milks should not be given to infants under six months old for a number of reasons, including that they contain more iron than young babies need as well as sucrose, glucose and other non-milk sugars. They also often contain higher amounts of micronutrients. There is no evidence that follow-on milks offer any nutritional advantage over whey-dominant milks, which is why they are not recommended by any major health organisations. The advice is that for formula-fed infants there is no need to change from first milk to follow-on milk after six months of age. Goodnight milks are also a thing. These are formula milks that contain follow-on milk and cereal (e.g. rice flakes, buckwheat flakes). They claim to help babies to sleep and are aimed at young children over six months of age. However, there is no independent evidence to show that they do indeed help babies to settle at night.

Growing-up milk and toddler milks are also available and are particularly responsible for the large growth in the formula milk market. They are marketed as an alternative to cow milk for children over one year old but have been deemed, by the EFSA amongst others, to offer no additional value to a balanced diet. These milks all contain a lot more sugar than cow milk and, in many instances, less calcium. They may also contain flavourings, such as vanilla, possibly encouraging a preference for sweeter foods.

For infants who can't have cow milk, for whatever reason, there are alternatives as either soya-based milks or specialised formula. Specialised formula might be amino-acid-based, rice-based, or protein hydrolysate formulas that contain protein (either whey or casein) that has been broken into smaller sizes than found in cow or soya-based milks. These milks have been developed to meet a range of specific needs and should only be used under the guidance of a health professional. The different classes of milk (i.e. cow-milk-based, soya-based and specialised formula) vary in their nutritional content, amount of calories, digestibility, taste and cost. Soya-based formula milks contain more non-milk sugars than regular formula milk and are considered more likely to result in tooth decay. Goat-milk-based formula milks are also available but are no less likely to cause allergies. Goat milk contains similar proteins to cow milk so is unsuitable for babies with cow milk allergy. There are also special formula milks for infants experiencing reflux, that contain additional thickening ingredients, such as corn starch and carob bean gum, to help the milk stay in the stomach rather than come back up so easily. There is some evidence that these may be of help in formula-fed babies with persistent and distressing reflux but they are not widely recommended for general use.

Is breast milk better than formula milk?
This is a question that has constantly been asked ever since the emergence of formula milks. At any one time, someone somewhere in the world is writing about this issue in the press or online, or talking about it on TV or at an event. Some may be promoting the 'breast is best' message, whereas others may be trying to combat guilt in mothers who choose to formula-feed. The whys and hows of infant feeding is for others to debate but I'm going to try and avoid the politics here and just take a look at what the science says.

Breastfeeding has been shown across many studies to help protect health and contribute to development. Scientists have found that breastfeeding is associated with a lower risk of diarrhoea, ear infections, respiratory infections, dental malocclusions, sudden infant deaths, diabetes (both Type 1 and Type 2) and obesity, for example. In addition, breastfed individuals do better on intelligence tests, with a higher average IQ by 3 points, improved academic performance (in some studies) and increased adult earnings. There are also proven health benefits of breastfeeding for the mother, with a reduced risk of developing breast cancer and ovarian cancer. There have been other claims made in favour of breastmilk but not all are substantiated by robust scientific studies. For instance, breast milk has not been sufficiently proven to protect against asthma, eczema or food allergies, and there's no good evidence to confirm a positive effect on blood pressure or cholesterol.

This information largely comes from observational studies but it doesn't explain how breast milk is conferring such benefits. There is still a lot to learn, but the diverse bioactive components it contains are clearly working to protect the young baby from disease and lay down healthy foundations for their long-term development. For example, breast milk contains many friendly microbes. Research into the role of the natural microbes in breast milk and how they affect the infant has been rapidly expanding. Early indications point towards breast milk being a vital factor in the development of a baby's gut microbiome, establishing healthy colonies and priming the infant's immune system, amongst other roles. Some formula milks specifically include friendly bacteria, as probiotics, in an effort to play a similar role but until we know much more about the specific microbes in breast milk and how they work, it is unlikely that formula milks will be able to mimic the same effects.

Leading health experts highlight that, in some countries, not breastfeeding has significant long-term negative effects on the health, nutrition and development of children and on women's health. One large and significant review of the evidence available from hundreds of separate studies, published in the *Lancet*, estimated that if a near universal level of breastfeeding was achieved globally, it could save over eight hundred thousand babies' lives. That is a huge projection but we really need to put all of this into context. Many of the benefits are greater in places where there are high rates of serious infectious disease, low levels of hygiene and poor nutrition. As well as conferring immunity benefits within the milk, breastfeeding helps to avoid contamination from dirty water sources and bottles, for example. For the many people who live in hygienic conditions in high income countries, the short-term health risks from formula feeding are much lower although stomach and ear infections, for example, are still relatively common. But even in such countries, the long-term benefits of breastfeeding, such as a reduced risk of dental malocclusions, diabetes, obesity and improved intelligence, are significant. It's worth noting that some of the advantages conferred by breastfeeding are not necessarily from the milk itself but potentially from other social factors and behaviours involved.

In theory, breast milk is the perfect food for a baby but in practice some babies may come to some harm from it. For example, a slightly higher rate of dental decay has been found in babies breastfed for one year or more and iron deficiency is more common in breastfed babies. A small number of babies also develop reactions to their mother's breast milk. It may not be a response to the milk itself but instead to something that is passing through into the milk from either the mother's diet or her exposure to particular compounds. The presence of proteins from dairy or soy, for instance, can cause problems for some infants, leading to a need to switch to special formula milk, and there may be other nutrients causing a problem. These can be difficult to identify. There are also other potential contaminants of breast milk that can cause harm, namely alcohol, tobacco and drugs (whether legal or illegal). In principle, anything a mother consumes may eventually find its way into her breast milk.

Infants can't process alcohol in the way adults can so alcohol in breast milk may affect them quite acutely, but at low levels this is unlikely to do much harm. At higher levels, however, the mother's alcohol consumption can affect the sleeping and eating patterns of babies as well as impact upon their early development. The amount of alcohol in breast milk depends not only on the amount consumed by the mother but also the time between its consumption and when the breast milk is expressed. For those of you who are breast-feeding and still like to have an alcoholic drink, you may have heard about 'pumping and dumping' – where you express the breast milk after consuming alcohol and discard it, in order to maintain milk production while avoiding exposing the baby to the alcohol in the milk. (I certainly know a few people who did this when they had their babies.) Unfortunately, this is not an effective way to reduce the alcohol content of the breast milk. Time is the key. Breast milk doesn't store alcohol up. The alcohol in breast milk works pretty much in the same way as that in the bloodstream, so just as the alcohol level in the bloodstream decreases with time, so it does in the breast milk. Breast milk continues to contain alcohol for as long as there is alcohol in the mother's bloodstream, so the more alcohol consumed, the longer alcohol will be present. In general, however, only a small proportion of alcohol passes from the mother into her breast milk.

Nicotine and other harmful chemicals found in tobacco can be passed into breast milk. While some medications may pose a risk in breast milk and can accumulate to produce toxic effects, most do not pose a risk at low levels for otherwise healthy babies. Herbal products are increasingly used by lactating mothers. While some are known to be potentially harmful, including kava and yohimbe, there is a lack of safety data for many of the most commonly used, including chamomile, black cohosh, St John's wort, echinacea, ginseng, ginko and valerian. Adverse effects in breastfeeding infants have been reported for some commonly used herbal products and a lot more research is required to understand how they work and their impact on infant health. Various illicit drugs have been found to get into breast milk and have a negative effect on babies' health, as well as treatments used to combat substance abuse. There is a huge range of pharmaceutical, herbal and illicit substances that work in different ways so women are advised always to consult with their doctor regarding any potential impact on breast milk.

Environmental contaminants, like pesticides and heavy metals, can also find their way into breast milk, depending on the exposure of the mother to such substances. This is why breastfeeding mothers are advised to limit their exposure to mercury in the diet, for instance, with certain fish being the main source. This may sound alarming but you need to remember that heavy metals and other environmental contaminants have been found in formula milks too. It has been concluded that, in most cases, the levels of such contaminants present in breast milk or formula milk are very low and are unlikely to pose harm to a baby.

Taking a look at additional potential harms from formula milk, there is obviously the case that some babies are allergic to cow milk or soya milk. This necessitates the use of specialised milk. Formula milk also has a significant potential for contamination during its preparation and storage. Formula-fed babies can become ill if their milk is not prepared safely. While relatively low-level contamination, for example from incomplete cleaning of bottles, may lead to an upset stomach (although this can be serious for vulnerable babies), there have been a number of very significant contamination scandals. In 2017, dozens of babies across the world fell ill after consuming formula milk contaminated with *Salmonella*. The firm behind the milk, Lactalis (one of the world's largest dairy firms), had to recall twelve million tins of the product from eighty-three countries due to the contamination. Worse still, formula milk contaminated with melamine (used in plastic and fertiliser production) led to the deaths of six babies and illnesses, some very serious and long-lasting, in over three hundred thousand more babies in China in 2008. Although you might think this was the result of a one-off mishap at a single factory, it turned out that twenty-two companies were involved in this tainted milk scandal and this has resulted in a local deep mistrust of Chinese-made formula milk.

The scope for human error is significant. Not only can formula milk become contaminated, the wrong amount might be used, or delivered at an unsafe temperature, or air can be introduced during feeding etc. Formula milk is commonly associated with gas, constipation and gastrointestinal discomfort as a result. As noted earlier, formula-fed infants are more likely than breastfed infants to develop a range of infections and other health complaints. There is also a clear environmental impact from the manufacture, packaging, distribution and use of formula milks and their associated products (e.g. bottles, sterilisers and other accessories) versus breastfeeding. In addition, a lot of infant formula contains palm oil. There have been many high-profile campaigns raising public consciousness about the devastating environmental impact of the palm oil industry. Palm oil plantations have led to huge deforestation, particularly in Indonesia, and the sharp decline of critically endangered species.

So, in answer to the question of whether breast milk is better than formula milk, I think that we have to say that on average as a foodstuff yes it is. With its complex composition of nutrients and bioactive compounds, breast milk is unlikely to be matched by formula milk for a long time. In high income countries where there are good hygiene practices and high-quality formula milks available, formula milk offers a safe, nutritious alternative. It provides the nutrients needed for a baby to grow and develop but as a whole it doesn't quite confer the same health benefits and protections that breast milk can potentially offer.

Animal milks

Even though consumption has been decreasing, animal milks still constitute a major part of the diet for most people. When I talk about animal milks, I'm mainly focusing on milk from cows, which is by far the most commonly produced and consumed. There are other animal milks that are also frequently drunk in certain parts of the world, including milk from goats, sheep, horses, donkeys, camels, buffalo and yak.

What is in animal milks?

The colour, flavour and composition of milk is dependent on a range of factors, such as the animal species it comes from, variation of breed, its age and diet, stage of lactation, individual and herd variations, as well as farming practices and seasonal and environ-mental variations. In general, animal milk is comprised of water, minute fat globules, proteins, milk sugar (lactose), minerals and trace amounts of other molecules, including vitamins (particularly B vitamins), enzymes, pigments and gases. By far the largest constitu-ent of milk is water. The water content of milk from different animal species ranges from 83 per cent in yaks to 91 per cent in donkeys. Cow milk contains about 87 per cent water. Buffalo and yak milks have a very high fat content, around twice as much as that in cow milk, and higher protein content. Yak milk is sweet and fragrant and is usually consumed raw in cups of tea by herding families. Camel milk is similar in composition to cow milk but is a little saltier and contains three times as much vitamin C. It is rich in unsaturated fatty acids and B vitamins and consumed either raw or fermented. Milk from horses and donkeys is relatively low in fat and protein but rich in lactose. It is usually fermented before consumption. Goat milk is actually very similar to cow milk but sheep milk, which you might think would be the same as goat milk, has higher fat and protein than goat and cow milk (making it particularly good for producing rich, thick Greek-style yoghurt). It also has more lactose than in cow, buffalo or goat milk. One study found that goat milk was higher in vitamin B12 (22 per cent higher) and folic acid (11 per cent higher) than cow milk, although other research has found that goat milk has no nutritional advantage over cow milk. Sheep milk has been found to contain about five times the amount of vitamin C and a higher amount of other vitamins also. That's a summary of what is in natural milk but commercially processed milk, particu-larly cow milk, may also be enriched with additional nutrients such as vitamins A and D (known as fortified milk).

How is milk processed?
Here, I'm describing the everyday cow milk most of us are familiar with. Raw milk refers to milk that has not undergone

pasteurisation and just gets bottled up as it is, but most of what we drink is processed. After the cows have been milked, the milk is stored in cold tanks and transported from the farm to the dairy for processing. Here, the milk is pasteurised to reduce the amount of potentially harmful bacteria, by heating it up very quickly then cooling it down again. The milk is separated into its cream component from its liquid component and then it is 'standardised'. This is when the liquid and cream components are recombined but to exact fat requirements for different milks, e.g. whole, semi-skimmed or skimmed. A lot of milk is also homogenised. As milk contains fat globules of a range of sizes, if left to set the larger globules rise to the top and form a cream layer. Homogenisation forces milk through a hole at high pressure to break down the larger fat globules and disperse the globules throughout the milk to give it a more even consistency. Filtered milk goes through an additional process. As its name suggests, it is filtered to remove further bacteria that have the potential to sour the milk and, as a result, extends its shelf life.

Talking of shelf life, have you ever been camping? Picture the scene . . . you're waking up in a fresh, dewy field, listening to the early morning birdsong and full of excitement for the day ahead only to be faced with depressingly warm and slightly funny-tasting milk in your cornflakes. Sound familiar? You've just been subjected to what I always think of as back-up milk: UHT milk, the kind of milk you keep in your cupboard in case you run out of the fresh stuff. UHT stands for 'ultrahigh temperature' or 'ultraheat treatment' and it refers to the processing that the milk has undergone. This milk is heated to around double the temperature of normal pasteurisation (about 140°C versus 70°C) for a few seconds. As fresh milk goes off very quickly, the purpose of UHT processing is to kill or inactivate all micro-organisms that would be likely to spoil the milk during transportation or storage. In effect, the milk is sterilised and this extends its shelf life from a few days to months and it no longer requires refrigeration (whilst unopened[4]). The use of UHT milk varies across different countries. While in the UK and the US we may consider UHT milk as a bit of a back-up and use it rarely, it is very commonly used in countries such as Belgium, Germany, France and Spain where it is often the main source of milk.

Lactose intolerance is a common malady where sufferers are unable to digest lactose – the sugar found in milk and other dairy products – properly. The body makes an enzyme called lactase to help digest lactose but if it doesn't make enough of it, you may not be able to digest lactose fully. Flatulence, bloating, stomach cramps, diarrhoea and nausea . . . just some of the symptoms experienced by people who are lactose-intolerant and who find regular milk difficult to digest. This may sound familiar to you or someone you know, but luckily there are dairy products that have been altered to make them easier to digest for people with this problem. The idea is that they can still benefit from the nutrients of milk without being hampered by the presence of lactose. This is, of course, lactose-free milk and it is made by a number of different methods. The first is by breaking down the lactose, by adding the lactase enzyme that converts it into the simple sugars glucose and galactose, which are then easily absorbed into the bloodstream for use as energy. These can make lactose-free milk taste a little sweeter than regular milk as simple sugars are sweeter than more complex sugars like lactose. The milk is then UHT treated to deactivate the enzyme and this also extends its shelf life. Another method is to filter out the lactose from the milk, which also removes some of the sugar as a result.

A fairly recent development in milk products is A2 milk. It is marketed as a healthier version of regular milk, and is particularly popular in Australasia and China, but what is it? Regular milk

---

4 Once opened, the milk is no longer sterile and is again exposed to microorganisms that could lead to spoilage.

contains the milk proteins A1 and A2 beta caseins. The proportions of each protein in your milk vary with the breed of cow it came from but it is typically around 50:50. A2 milk, however, just contains the A2 beta casein. Unlike some of the other types of milk I've mentioned, this isn't regular milk that has somehow been altered to produce a different form of the milk; it comes directly out of the cow as A2 milk. The company behind this product identify cows that naturally produce milk that is free of A1 beta casein, via genetic testing, and use exclusively those cows to source their milk. Sounds interesting – but what's the point of this? Proponents argue that A1 beta casein in milk can cause indigestion and gastrointestinal discomfort for many, that lactose-free milk can't solve. It is claimed that the A2 protein more closely matches human milk so doesn't lead to such problems. I'll come on to the evidence for these claims later.

For milk to be certified as organic it must satisfy a number of specific criteria. The cows must be free range (i.e. be at pasture where possible, for more than two hundred days on average) and graze on pasture where fewer pesticides and no artificial fertilisers are used. They are fed a grass-rich diet, free from genetically modified feed, and are not subjected to the routine use of antibiotics. There should be high standards of animal welfare and the farming system should work with nature to promote broader wildlife benefits and biodiversity. Although none of that relates to the processing of milk as such, organic cows are typically not pushed to maximise their milk yields where some other cows can be. Yields in organic milk production are about 20 per cent lower, on average, than in intensive production.

What's the difference between 1 per cent milk, skimmed, semi-skimmed, whole, full-cream, evaporated, condensed and fermented milks?
I don't know about you, but I find the language of milk confusing, and especially so when I travel abroad. So I found it useful to summarise the main terms here. In the UK, the terms 1 per cent, skimmed,

semi-skimmed, whole and full-cream simply relate to the proportion of fat in the milk: 1 per cent milk has a fat content of 1g per 100g, skimmed milk contains a maximum of 0.3g fat per 100g, semi-skimmed milk has a fat content between 1.5 and 1.8g per 100g and whole milk has a minimum fat content of 3.5g per 100g. Other countries use different terms and slightly different fat proportions such as reduced-fat milk (2 per cent fat), low-fat milk (1 per cent) and even fat-free (containing no more than 0.2 per cent fat in some countries and no more than 0.15 per cent in others). Full-cream milk is the same as whole milk. The cows from the Channel Islands, Jersey and Guernsey, are known for producing particularly rich and creamy milk and this can be bought as another category altogether. As the categories are all very specific, this is why the fat content is removed during processing and then added back in very precise amounts to ensure consistent fat proportions in the various types of milk.

Although not technically drinks, it's worth explaining the difference between condensed milk and evaporated milk (mostly because people keep asking me this question). Both condensed milk and evaporated milk are produced by removing some water (around 60 per cent) from whole or skimmed milk via evaporation. The heat during the evaporation process imparts a slightly caramelised flavour and light beige colour to the milk. The main difference is that condensed milk is then sweetened with sugar. Evaporated milk endures slightly more processing than condensed milk as it requires intensive heat sterilisation to preserve the milk, whereas the addition of sugar helps preserve condensed milk because sugar naturally inhibits microbial growth. Some countries refer to evaporated milk as unsweetened condensed milk.

Fermented milks have been around for thousands of years in many countries but are also now becoming increasingly popular in others, such as the UK, US and Australia. These milks are made by specifically adding micro-organisms to milk to achieve a particular level of acidity. This microbial culture converts some of the lactose in the milk into lactic acid, resulting in the formation of substances, such as carbon dioxide, acetic acid, diacetyl, acetaldehyde and ethyl alcohol, that give the products their characteristic taste and aroma. There are numerous versions of fermented milks around the world, including *kefir* (fermented cow or goat milk, hailing from West Asia/ Eastern Europe), *filmjolk* (fermented cow milk favoured in Sweden), *kumis* (fermented horse milk, from Central Asia and Columbia), *kule naoto* (traditional fermented cow milk of the Maasai in Kenya), *tarag* (traditional Mongolian fermented milk of cows, yaks, goats or camels), *airag* (traditional Mongolian fermented horse milk) and Yakult (a commercial fermented cow milk drink). Although it isn't typically referred to as a fermented milk, buttermilk also falls within this category. In the past, buttermilk was the liquid residue left over from the butter-churning process, but these days it is made by adding lactic acid bacteria to pasteurised low fat milk. This gives it a tangy flavour and slightly thickened texture. The differences between these fermented milks, beyond the type of animal milk used, are due to differences in how they are made, the microbial cultures used and whether they start with raw or pasteurised milk, amongst others. Fermentation helps preserve the milk to some extent, lengthening its shelf life.

Milk and health
Two young fans of Liverpool Football Club are chatting in a kitchen. One says to the other that Liverpool player Ian Rush had told him that if he didn't drink milk he would end up playing for Accrington Stanley. The other then asks '*Accrington Stanley, who are they?*', to which the first boy replies '*Exactly*'! This television advertisement from 1989 by the UK's Milk Marketing Board was on our screens for several years and many of us remember it well. If we didn't drink our milk, we wouldn't reach our potential, it would seem. Over the years, there have been lots of adverts encouraging us to drink more milk, but as time has gone on, dairy has had an increasingly fluctuating

relationship with public perceptions about whether or not it is healthy. Stories about links with cancer, obesity, heart health and osteoporosis, for instance, have filled the news and led to scepticism about dairy. So where are we now with the science behind all this?

Researchers have undertaken large reviews of the scientific data on milk consumption and have generally found no significant harms. In fact, for the most part, the studies point towards a potentially protective role of milk consumption against cardiovascular disease, childhood obesity, stroke and some cancers (e.g. colorectal and breast cancers), as well as benefits for bone mineral density. In other areas of health that have been investigated, no strong link has been found (i.e. no specific benefit or harm), including for bone fractures, adult obesity and a number of other cancers. There is some limited evidence that dairy consumption may be associated with an increased risk of prostate cancer but the link is not yet clear. Quite a few studies into the effects of milk have been carried out in conjunction with the dairy industry so are not totally without potential bias, however other studies with no such industry links have in fact found similar results.

One area of concern for some is the saturated fat content of animal milk. Eating a diet high in saturated fats is associated with raised levels of 'bad' cholesterol (non-HDL[5] cholesterol) and this is linked with cardiovascular disease. In general, the advice is that we should reduce the saturated fats in our diet. However, researchers looking into this have found that not all saturated fats are equal, with different fatty acids seeming to have different effects on the body. Furthermore, alongside saturated fats, milk contributes a higher intake of other significant nutrients that appear to have a beneficial effect on cardiovascular health. In effect, it may be the case that while one component of milk has the potential to impact our cholesterol levels, other components may be working in a

5 High-density lipoproteins.

cardio-protective fashion. So, when consumed as a whole, milk seems to be beneficial, rather than harmful, for our hearts. Having said all this, milk isn't actually that high in fat, when compared with other foods in our diet. The relevance of the saturated fat content in milk will depend, amongst other things, on what type you drink (i.e. full or low fat), how much you drink and how regularly you drink it.

Other than less fat and fewer calories, there is little difference in nutrient content between skimmed, semi-skimmed and whole milk. The exception is vitamin A content, which increases as the proportion of fat increases (e.g. whole milk contains around twice the amount of vitamin A in semi-skimmed milk but around fifty times the amount that's in skimmed milk). This is because it is found within the fat in milk. You often hear that milk is a good source of vitamin D. One reason for this is that in some countries, such as the US, vitamin D is added to the milk. However, milk is not typically fortified in the UK and the naturally occurring amount of vitamin D in milk is negligible. A number of nutrition researchers and clinicians recently called for milk in Europe to be fortified with vitamin D as standard as vitamin D deficiency is so widespread. They believe that fortifying this commonly consumed product could significantly improve public health and slash healthcare costs by billions of euros. We'll have to wait and see if any action is taken to fortify milk on such a large scale and whether the dramatic public health claims pan out as a result. In the meantime, individual manufacturers are producing their own fortified milks and hoping to capitalise on the public's escalating interest in health.

One of the key roles of vitamin D is to help the body absorb calcium. When asked to name something healthy about milk, most people would probably mention its calcium content. Calcium is essential for our health, not only for growing and maintaining a healthy skeleton but also for a host of other functions, such as blood clotting, muscle contractions, sending nerve signals and regulating our heartbeat. There's no denying that milk is a good and convenient source of calcium but we can also get it from elsewhere in our diet, like green leafy vegetables, nuts, tofu and (increasingly) fortified foods.

Organic milk

Some researchers have pointed towards higher nutritional benefits from organic milk, namely beneficial omega-3 fatty acids and vitamin E. However, while the study behind these claims took into account 196 research papers, their findings were blown out of proportion according to many experts in the field. The difference found in omega-3 fatty acids in organic milk actually amounted to only a 1.5–2 per cent increase in our diet overall and would be unlikely to represent a nutritional or health benefit, and milk is a poor source of vitamin E so the increased level found in organic milk was also not nutritionally relevant. What is widely known about organic milk is that it is significantly lower (up to 40 per cent lower) in iodine than regular milk, which is a problem because milk and dairy products are the primary sources of iodine for many people. The reason for this difference is that non-organic cows receive iodine-supplemented animal feeds in conventional farming, as well as exposure to contamination from the use of iodine-based products in sanitisation practices. The body needs iodine to make thyroid hormones, which are essential for growth, regulating metabolism and foetal development. As your body doesn't manufacture iodine, you need to get it from your diet so your choice of milk might make a difference if you have low levels or don't consume it from any other sources (e.g. fish). This could be particularly relevant for pregnant women who require additional iodine in their diet to support the healthy development of their baby. There have been calls to supplement organic cows' diets with natural sources of iodine (such as seaweed) and some farmers are now taking steps to address this discrepancy in iodine levels.

Raw milk

Milk is an ideal medium for microbiological growth, either from the environment or the animal itself. Milk can contain a range of potentially harmful micro-organisms, such as *Salmonella, E. Coli, Campylobacter, Listeria, Staphylococcus aureus, Clostridium botulinum, Brucella* and a number of harmful moulds, given the right conditions. In addition, zoonosis infections (diseases that may be transmitted between animals and humans) are not uncommon from milk and milk products, including tuberculosis, leptospirosis, listeriosis, brucellosis and salmonellosis. It is for these reasons that most milk is treated to avoid microbiological harm to the consumer. However, some people purposely choose to drink untreated milk. Fans of raw milk like the idea that it is a completely natural product made without humans messing about with it. Many believe it is rich in nutrients and crammed full of friendly bacteria to aid our guts, and some even claim it boosts the immune system and protects against allergies. But what does the science say? Well, raw milk is definitely full of bacteria but unfortunately there are a lot of unfriendly ones in there. Because it hasn't been pasteurised, raw milk may contain harmful bacteria that could cause food poisoning and other illnesses. My own mother, as a child, was infected with bovine tuberculosis from drinking unpasteurised milk. There are numerous documented instances of raw milk causing illness and due to this risk to public health, raw milk cannot be legally sold everywhere, only in certain countries and settings, and vulnerable groups (e.g. the elderly, small children, pregnant women, those with a compromised immune system) are advised not to consume it. Although many farms take good measures to limit potential microbial risks, studies have found that up to a third of raw milk samples contained pathogens, even when they were from healthy animals and from milk that appeared good quality. So why do people choose to take the risk?

Raw milk has indeed been found to contain microbes that exhibit probiotic properties, such as *Bifidobacteria* or *Lactobacillus acidophilus*. However, their levels should be low in milk, as they do not compete well with other naturally occurring milk bacteria, and in fact the presence of high amounts of *Bifidobacteria* in raw milk has been used as a possible indicator of faecal contamination. Their role in health has yet to be fully illuminated but some researchers highlight the potential for re-introducing these helpful bacteria into milk once it has been pasteurised to gain potential benefits while minimising potential risks. Others point out that when introducing probiotics into foods it is better to use strains originally isolated from human sources, not from cows. Until more is known about the use of probiotics in milk, and given the risk from harmful bacteria in raw milk is significant, it is suggested that we get our beneficial bacteria from other dietary sources instead.

It also seems true that pasteurisation slightly reduces the nutrient content of milk but these nutrients are more than made up for by our usual diets. For instance, vitamins B12 and C are decreased when the milk is pasteurised but the losses are minimal and milk doesn't actually provide a significant source of these vitamins in our diet anyway, so this is of little consequence. Vitamins D and K do not seem to significantly decrease through pasteurisation and although vitamin B2 (riboflavin) does decrease and milk offers a good source for it, we can still get plenty of this particular vitamin from our diet. Vitamin A levels have actually been found to increase after pasteurisation.

While supporters say that raw milk can help protect against childhood asthma and allergies, pointing to studies of people who've grown up in rural areas, there is no strong evidence to support the consumption of raw milk in relieving allergies. It is believed that raw milk contains natural proteins, antibodies and microbes that help to support a strong immune system and avoid allergies, and that these are destroyed by pasteurisation. However, the mechanisms underlying such an effect are unclear and there just isn't enough proof to support this claim. In short, the risks of drinking raw milk seem to far outweigh any possible benefits (for which there isn't much evidence).

UHT milk

At the other end of the scale from raw milk is UHT milk. It is a handy thing to have around but is it just as nutritious as regular pasteurised milk? UHT processing of milk can reduce its folate content by 20–30 per cent and it may lose a substantial amount more during storage. This is unfortunate as milk is typically a pretty good source of dietary folate. Vitamins B6, B12 and C are also particularly depleted in UHT milk versus pasteurised milk. In the UK, for example, UHT milk is something most of us only drink now and then when fresh milk is unavailable so we probably shouldn't worry about the issue of depleted nutrients too much. However, in many European countries it is the main source of milk so nutritional deficits in this form of milk may become more relevant.

Homogenised milk

The homogenisation of milk is something that worries/causes mild hysteria amongst some people. The concern is that the homogenisation process alters the structure of the milk and this may lead to an increased risk of cardiovascular disease and other illnesses. The issue raised is to do with an enzyme called xanthine oxidase (XO). Elevated XO in the body has been found to induce inflammation and has been linked with an increased risk of developing type 2 diabetes and cardiovascular disease. Milk naturally contains plenty of XO. When the fat globules are reduced during homogenisation, their surface area increases and they are no longer completely covered with the original membrane material. Instead, the surface sticks to a mix of proteins. In non-homogenised milk, XO is thought to be on the surface of the membrane surrounding the fat globules where digestive juices in our stomach can access it and break it down before it enters our body, but after homogenisation it is suggested that XO gets encapsulated by the new smaller fat molecules and is in some way protected from the digestion process and is absorbed intact into the blood stream. This apparently then goes on to damage cardiovascular tissue and clog up the arteries, leading to heart disease.

The problem is that this is all based on a decades-old hypothesis from one person (Dr Kurt Oster) that has since been discredited by many others. It was a theory, not based on findings from actual scientific trials. In fact, there is no robust evidence to demonstrate a relationship between homogenised milk and levels of XO in the body, nor a relationship between homogenised milk and cardiovascular disease, nor even that milk XO is even absorbed into the body – it has not been found that XO is absorbed from any dietary source, rendering the whole argument pointless. Yes, our bodies do contain XO but we're manufacturing it ourselves not consuming it (what causes levels to increase or decrease is another physiological matter entirely). Despite the theory being originally disproved way back in 1983, by researchers from the University of California at Davis (in the *American Journal of Clinical Nutrition*), and backed up by subsequent research that also debunked it, scepticism surrounding homogenised milk has persisted and a huge number of blogs and websites continue to demonise it. As with so many health theories, it is fascinating how people are willing to believe the *ideas* of one person yet ignore *actual scientific evidence* laid before them by so many others. People believe what they want to believe (although they probably wouldn't want to know that homogenisation is common in the manufacture of other milk-based products, such as chocolate milk, iced coffee drinks, ice cream and yoghurt, in case they have to eschew these tasty treats too).

Lactose-free milk
Lactose intolerance is thought to affect around 65–70 per cent of the world's population due to genetic differences between those who can and cannot tolerate lactose. It is more common in particular

ethnic groups (for example in people from East Asia) but in the UK the numbers are relatively low as it is estimated to affect only around 5 per cent of the population. We are all usually born with plenty of the lactase enzyme and the subsequent ability to digest lactose; only very rarely are babies born with lactose intolerance. We need sufficient lactase to digest breastmilk (which contains nearly 70 per cent more lactose than cow milk) but, for many people, lactase levels decline after weaning and when your diet becomes less reliant on milk and dairy products, around the age of two. In these individuals, lactose intolerance doesn't really show up clinically until after the age of about five. Back in the mists of human evolution, we all lost the ability to digest lactose after infancy but a genetic mutation in our relatively recent past (probably around 7,500 years ago) spread throughout Europe, in particular,[6] enabling a proportion of the population to maintain their ability to produce the lactase enzyme (known as lactase persistence) and adapt to drinking unfermented animal milks.

This all refers to primary lactase deficiency (i.e. it's in your genes and tends to run in families), but secondary lactase deficiency can arise at any age as a result of a problem occurring in your small intestine, whether from another health condition, surgery or medication. Milk free from lactose is a useful aid to sufferers and, other than a change in the sugars, is considered to contain the same nutrients as regular milk. These individuals (unless severely intolerant) are often advised not to cut out dairy products altogether to avoid missing out on vital nutrients in the diet. Many lactose intolerance sufferers can still enjoy other sources of dairy (e.g. cheese) without experiencing discomfort.

---

6 Separate genetic mutations promoting lactase persistence are also judged to have occurred elsewhere, in West Africa, the Middle East and South Asia. (Curry, A. (2013) 'Archaeology: The milk revolution.' https://www.nature.com/news/archaeology-the-milk-revolution-1.13471)

Lactose intolerance is not the same as having a milk allergy. While lactose intolerance means you lack the means to properly digest lactose, a milk allergy is an abnormal reaction of the body's immune system in response to milk consumption, particularly the proteins in the milk.[7] Symptoms range from mild to severe and may include a rash, wheezing, runny nose, stomach cramps and vomiting, and can even lead to anaphylaxis (an extreme, life-threatening reaction affecting the whole body). Milk allergy is estimated to affect between 2 and 7.5 per cent of infants under the age of one. Although it is a relatively common allergy in children, most will outgrow it. Strict avoidance of milk and dairy products is the only way to deal with this allergy. So, you can see that lactose intolerance is far more common than milk allergy. And, whilst we're on the subject, you may have read that goat milk is better for people who are lactose-intolerant due to its lower lactose levels. However, this simply isn't the case. Goat milk contains a very similar amount of lactose to cow milk and has been found to be no less allergenic.

A2 milk

Over the years, the claims regarding the benefits of A2 milk have changed. First, it was touted as being healthier than regular milk because while regular milk was associated with a higher risk of type 1 diabetes and heart disease, A2 milk was not. However, such a claim is unsubstantiated by research and in fact evidence shows that there is no such link between regular milk and these health conditions. In more recent years, the links with digestion have been promoted instead. In order to prove the benefit of A2 milk, you need to prove that the A1 beta casein is somehow harmful to health. However, studies looking into this very issue haven't so far been able to demonstrate definitively that it causes ill health in humans. A few studies have found links but others have not. Evidence is currently sparse

7 Milk allergy is also referred to as CMPA (cow's milk protein allergy).

and inconsistent and far more research is required to robustly validate a beneficial link between A2 milk and health. Despite scepticism about the science behind A2 milk, what started out as the brain child of a small New Zealand start-up company (A2 Milk Company) has now been jumped upon by some of the world's largest dairy businesses, including Fonterra Co-operative Group Ltd (the world's largest dairy exporter) and Nestlé SA. It is big business that is enjoying a surge of popularity and companies are lining up to capitalise on the trend. After all, they can charge a lot more for a bottle of A2 milk than regular milk and recoup significant profits.

Whey protein drinks
Talking of milk proteins, the whey protein portion of milk is used in drinks traditionally marketed to serious body builders looking to bulk up their lean muscle. These days, others are also giving it a go, such as recreational gym-goers and those trying to lose weight, as a quick way to get extra protein into their diet. Many claims over the benefit of whey protein have been made, including muscle building, feeling of fullness, body fat reduction, increased endurance and faster recovery after exercise from muscle fatigue. A few years ago, the EFSA decided to look into such claims and, in summary, rejected them all. They were not convinced there was strength of evidence to support them. More recent studies have found some effect of whey protein on muscles in people who undertake resistance exercise, although evidence of wider effects is still limited. A number of possible harms of whey protein drinks have also been mooted. The addition of sugars in some drinks increases the calories people would be ingesting, they may cause intestinal discomfort due to the milk-based protein being used, and, more worryingly, a report in 2018 found that many products contained contaminants such as heavy metals and pesticides. If you want to gain extra protein, some experts advise you find it elsewhere in your diet but if protein supplements are to be used, they should be used sparingly. In an entirely different

use, whey protein has been found to be useful in the nutritional support of malnourished individuals.

Fermented milks

Controlled cultures, fermentation temperature and time are key to developing edible fermented milks. It's not just a case of letting your pint of milk go off in the fridge. Fermented milks are thought to be easier to digest than regular milk, so much so that in some cultures fermented products are used as weaning foods for infants. One reason for this is that fermented milks have lower lactose levels than regular milk as the fermentation process breaks down some of the lactose, making it more digestible for some who don't tolerate milk well. A number of fermented milks have shown promise for our health, one of which is the increasingly popular kefir.

A key aspect of kefir that distinguishes it from other fermented milks is that a kefir grain is used in the fermentation process and it contains a large population of yeasts. Kefir grain is not actually grain at all but gelatinous beads, that look a bit like grain, containing a large diversity of bacteria and yeasts. This complex drink contains many components that have the potential to confer benefits, whether from the milk itself, or from the wide range of bacteria and yeast species (it's estimated that there's over three hundred different microbial species in kefir, although this varies widely). Kefir has long been claimed to offer health benefits but it's only in recent years that much scientific research has been trying to nail down whether these claims hold water. It seems that benefits associated with kefir are indeed emerging, from improved digestion and lactose tolerance, aiding the immune system, improving cholesterol metabolism and potentially having a role in alleviating allergies, as well as demonstrating anti-inflammatory and anti-cancer properties. However, many of the findings are from animal or lab studies. Robust clinical trials are now needed to demonstrate measurable effects in humans, as well as research to unpick the mechanisms underlying the effects of kefir, to what extent it is useful and who it may benefit, for instance. Currently, kefir is produced in many different ways across the world and the content varies hugely, so concomitant effects will likely also vary.

Research is also looking into potential health benefits of many other traditional fermented milks and the picture is similar to that of kefir. Emerging evidence suggests that these drinks may well be conferring some benefits but extensive trials are really needed to clarify what components are associated with particular effects, and in which drinks and populations. With the vast amount of microbial activity going on, there is equally the potential for harm. Interestingly, a small study of a Kenyan fermented milk, *mursik*, suggested that there may be a link between its regular consumption and oesophageal cancer, due to repeated exposure to carcinogenic levels of acetaldehyde, which is produced during the fermentation process. Further work will be necessary to confirm this but it does illustrate that not all fermented milks are equal. In fact, they are hugely diverse and, as you might imagine, with so many being traditionally made, there is a lack of regulation and guidance surrounding their safety, quality and use.

Microbial fermentation is an intricate business resulting in complex interactions between bacterial and yeast species and the milk that it is fermenting. With the staggering variety of fermented milk products in existence, the potential for health effects is significant but pinning down the exact components that are having the most impact is key. Understanding the most beneficial elements of fermented milks can then lead to the commercial development of products that are safe, with a consistent quality and composition and that are likely to exert the greatest effects. Manufacturers are trying to hit the sweet spot with their fermented milks but have yet to develop a conclusively, significantly beneficial product.

Unlike kefir and other traditional fermented milks, commercially produced probiotic yoghurt drinks typically only contain one or two species of bacteria. There is some evidence that drinking probiotic fermented milk containing *Bifidobacterium lactis* and lactic acid bacteria is associated with an improvement, albeit modest, in digestive discomfort in healthy adults, although this research was commissioned by Danone, a manufacturer of fermented milks, amongst other products. However, other researchers aren't so convinced. A review looking into the advertising claims regarding products containing *Lactobacillus casei* and *Bifidobacterium lactis* (such as Yakult and Actimel) concluded that there is insufficient evidence to support health claims made for these products, and other research has found that despite the presence of the friendly bacteria, if they actually make it through the stomach's acid environment into the intestine they may only benefit a small proportion of people, usually those who already have something wrong with them, and really need to be tailored to individual needs. We'll explore probiotics (and prebiotics) in more detail when focusing on wellness drinks, in Chapter 4, Cold drinks.

Antibiotics in milk

There have been claims that cow milk contains high levels of antibiotics that could be harmful for our health. Antibiotics are used in the dairy industry to treat common disorders in cattle, such as mastitis, foot problems and reproductive disorders. In the UK, antibiotic use in cows is actually much lower than in pigs or poultry and, unlike in some other countries, mass medication is not used frequently and usage is relatively low. Regulations in the UK, and some other countries, state that milk from cows that receive antibiotics should not enter the food chain and there is a strict withdrawal period for the milk (i.e. the milk from these animals is disposed of for a period of time). Milk is routinely tested for the presence of antibiotics. If the tests are positive, the milk cannot be sold and farmers may be fined or have their licence to produce milk removed. Despite the regulations and checks and balances in place, a number of studies from different countries that have tested samples have found residues of medicines in milk due for human consumption. They were only found in a fraction of the samples and the presence of antibiotics was generally at a very low level; unlikely to cause direct harm. However, it is possible that there could be a long-term effect. We just don't have the evidence either way. To put this in context, you may remember that I previously discussed how our tap water is also full of medicines residues, so milk substitutes made up with water also have the potential to contain them. The presence of hormones in milk also worries people but, while they are used in other countries, the use of hormonal growth promoters for livestock is banned in the UK.

Before we move on to plant milks, we need to mention one of the major criticisms of the dairy industry: its impact on the environment. Dairy is a large, global industry with an estimated 270 million dairy cows in the world. The industry impacts the environment in different ways, such as the production of greenhouse gas emissions and contamination of local water resources, and unsustainable farming practices, for both the cows and their feed, can lead to the loss of land. There's also the significant water and electricity usage required to keep and feed the animals. The greenhouse emissions, land use and water use are all much greater in the production of dairy milk than plant milks. According to some climate scientists, avoiding meat and dairy is the single biggest way to reduce our environmental impact on the planet.

Plant milks

Plant milk is a booming industry. In the UK alone, plant milk sales are predicted to rise to $400 million by 2021. Almond milk, soya milk and coconut milk are the best sellers, although other plant milks are gaining traction in the marketplace. Once the preserve of vegans and the dairy-intolerant, plant-based milks have gained enormous popularity among the rest of the population. Interestingly, statistics show that the majority of people who buy these milk

---

Night-time milk

We've all heard that drinking milk, particularly warm milk, at bedtime can help you drop off to sleep. The evidence to prove this actually works is pretty thin (see *What are malted milk drinks?* for more on this) but did you know that there is actually something called 'night-time milk'? This isn't referring to your usual milk but to milk that has been collected from cows milked at night. This is because they apparently have an increased level of melatonin in their milk. Melatonin is a hormone naturally produced in your body that plays a role in sleep and its levels increase and decrease depending on the time of day. It increases in the evening and night-time (when it is dark), decreasing again when it gets light, and works in conjunction with other bodily mechanisms to help your body to prepare for sleep. You may feel calmer and less alert and, as such, melatonin sleep aids are very popular. While it is thought that melatonin-enriched milk aids sleep, and is marketed as having such an effect, the evidence so far from scientific studies about its effects is sparse and inconsistent.

---

alternatives also buy regular dairy milk. Many plant milks have been around for centuries and traditionally used but in recent decades have been commercially exploited for the mass market. So what is fuelling this interest in milk alternatives? A variety of motives underlie this branching out, away from regular milk, including fears surrounding dairy consumption, a feeling that plant milks are somehow more wholesome, a generic desire to cut down on animal products, increasing discomfort about industrial farming practices[8] and

8 There are significant ethical issues surrounding dairy farming that there's no space to cover here.

the environmental impact behind dairy milks, and just wanting to trying something new. And in some developing countries animal milk is expensive so plant milks are a more affordable option. There are now myriad choices in the world of milk alternatives and it's tempting to assume they are all much of a muchness, but there are key differences, so let's take a look at these.

What are nut-based[9] milks?

Almond milk is the most popular plant milk on the market, having overtaken soya milk a few years ago. While milk alternatives in the supermarket may be dominated by almond milk, there are many other nut milks available (cashew nut, macadamia nut, hazelnut, pecan nut, walnut, pistachio nut etc). When you hear the process of how nut milks are produced, you'll realise why almost any type of nut can be exploited in this way. Nut milks are typically made by soaking, draining and rinsing whole nuts, then blending the nuts with water and straining out the pulp. The resulting liquid is the 'milk' and is chilled. This is the basic method that you can try at home (there are loads of blogs and videos online about how to do it), but commercially made nut milks typically add extra ingredients. Some of these help improve the nutritional worth of the product, such as the addition of calcium and vitamins, but others aim to alter the mouthfeel, such as thickeners (e.g. locust bean gum, carra-geenan, rice flour) and emulsifiers (e.g. sunflower oil), and the taste, such as sugar, salt and flavourings.

What are legume-based milks?

Soya (also known as soy) milk is made from soya beans or soya protein. Soya beans are soaked, crushed and cooked and then the liquid is extracted to produce the 'milk'. Commercial versions of

9 Technically, some of these are not nuts but I'm treating them as nuts because that's how most people think of them.

soya milk also contain ingredients such as apple extract, acidity regulators, salt, stabilisers, thickeners and sweeteners, and are fortified with added calcium and vitamins. The use of soya to produce a milk goes back hundreds of years in China but it is only in relatively recent times that its popularity truly expanded beyond East Asia, from the early twentieth century onwards. Soya milk was one of the first widely available dairy milk alternatives and was at one time the go-to for vegans and the lactose-intolerant, but has since declined in popularity. Health concerns about possible harms from soya arose and led consumers to seek alternatives.

Another legume-based milk is pea milk. It is not what you may be thinking – some sort of sludgy, green product of the petit pois – but in fact a drink made from yellow split peas. The split peas are milled into flour and the pea protein is separated from the starch and fibre. This protein is then blended with water and other ingredients to make the milk. These other ingredients include sunflower oil, sugar, algal oil, salt, thickeners, flavourings, calcium and vitamins.

What are cereal and seed-based milks?
Cereal milks (e.g. rice milk, oat milk, spelt milk) are typically made by soaking the chosen cereal, then finely blending it and adding water. Other ingredients are also often added, including sunflower oil, salt, acidity regulator, fibre, stabilisers, thickeners and calcium and vitamins. Seed milks (e.g. flax milk, hemp milk, sunflower milk) are made in much the same way as cereal milk although they are not always soaked first. In case you're wondering, hemp milk is made from the seeds of the hemp plant, a variety of cannabis plant. These seeds only contain a trace amount of THC, the psychoactive compound found in marijuana, and low-THC varieties are commonly used in the food industry in many countries, including the UK, US, Canada, Germany, the Netherlands, Belgium, Switzerland and Austria. These have none of the psychoactive effects associated with illicit cannabis.

Like cereal milks, seed milks also contain a number of added ingredients common to dairy alternatives. Some also contain pea protein to boost the nutritional worth of the drink.

What is in coconut milk?
Here, I'm referring to the carton of drink rather than the canned form you might use in a curry. It is made from a blend of coconut cream and coconut water plus water, and added ingredients, including sugar, grape juice, emulsifier, thickeners, stabilisers, salt, flavourings, calcium (sometimes other minerals, including magnesium and zinc) and vitamins.

Although I've listed the ingredients that might be contained in commercial plant milks, they are not in all of them. Some of the more expensive brands contain only a few ingredients and unsweetened versions of the drinks are widely available. Commercial plant milks are typically pasteurised, homogenised and sterilised to ensure safe products that are of consistent and aesthetically acceptable quality, as well as having an adequate shelf life.

In terms of taste, there is quite a difference between plant milks. Nutty, toasty, salty – all words that have been used to describe nut milks. Some consider cashew milk to taste less nutty and more creamy. Hemp milk is also described as having a nutty hint although some also mention its oiliness. Taste testers have decided that flax milk is pretty neutral in taste and texture. Rice milk ranges in taste but some of the biggest brands are thought to taste closer to dairy milk than some other plant milks, although a little more watery. Soya milk is described by some as having a rich, chalky taste and texture, although others mention its creaminess. One of the reasons that soya milk became less popular and almond milk surpassed it, is that people have complained about a 'beany' taste in soya and prefer the taste of almond milk. Like rice milk, it is considered to be closer to dairy milk and have a more rounded flavour than other diary alternatives. Oat milk is rather grey in colour, with a light, neutral flavour. Pea milk, in contrast, is very creamy and rich, comparing more in mouthfeel with whole milk. Quite a few brands of plant milks are found, by taste-testers, to be thin and watery, with a pretty neutral taste. This may be preferable as a plain drink or on cereal, for instance, but when added to strong drinks like coffee they may have little effect on the overall taste and so are probably a bit pointless in this regard (if the aim of milk in your coffee is to reduce the bitter flavour of black coffee, for instance). Coconut milk, on the other hand, is relatively sweet and creamy, which can overwhelm the taste of coffee (and certainly tea) and is felt by some to be too rich to use as an everyday dairy substitute. If taste matters to you, you may need to choose your milk alternative wisely!

Milk alternatives and health
'*Naturally lactose free*', '*Gluten free*', '*Low in saturated fat*' – all claims made on plant milks. A lot of claims focus on what the products don't contain, because all these statements are true and can't be contested. Quite frankly, you could say all of this about a glass of water. But what about what *is* in them? A few boast of being fortified with calcium and vitamins and some point to their naturalness . . . '*Plant powered goodness*', claims one product – what does that even mean?! Whatever the marketing hype, plant-based milk alternatives are often positioned as being healthier than regular dairy milks but while there has been considerable research into the effects of cow milk on health, there is a distinct lack of research into the direct and measurable effects of plant milks. Despite this paucity of evidence, we can still take a look at what they contain and consider some of the claims associated with them.

While the individual plants that form the basis of alternative milks have great nutritional profiles, this doesn't necessarily reflect on the quality of the milks themselves. Claims are made about the benefits of the flag-bearing ingredients (such as oats or nuts) but these claims usually refer to that single ingredient eaten alone, not as part of a combination of nutrients. To assess the evidence of any effects in real terms, in other words of what you are actually consuming, you need to investigate the whole drink itself as nutrients may interact with each other and the manufacturing processes may also have an impact on the end product. In fact, very little of the plants makes it into the milks. For example, while many nuts are packed with protein, healthy oils and fibre, and have been associated with a lower risk of heart disease and diabetes, you may be surprised to learn that the most popular brands of nut milks typically only contain around 1 to 2.5 per cent of nut content, and are mostly water. A few more expensive brands include around 5 to 6 per cent nut content. At these kinds of levels, you can't really expect many of the benefits of these base ingredients to be realised, unless you drank gallons of it every day. Cereal milks are a little better with rice and oat milks containing between 10 and 17 per cent of their main plant ingredient, but water is still the major component of the drink. The high fibre content is a selling point for oat milk (it contains about 2g per glass), which is a useful contribution to your daily intake (a recommended 30g per day). Oat milk contains the soluble fibre beta glucan, which has been found to help maintain normal cholesterol levels. One glass of oat milk contains about one third of the daily requirement for beta glucan.

One of the attractions of plant milks is that many are lower in calories than whole dairy milk. Rice, sesame and hazelnut milks have a similar calorie content to regular semi-skimmed cow milk, and hemp, oat and coconut milk have around the same amount as skimmed milk. Other plants milks have even lower calorie contents. Most of the calories come from carbohydrates. Rice milk has a similar or higher sugar content to regular cow milk, but other (unsweetened) plant milks contain less sugar. The addition of sugar, sweetener or other flavourings is common in plant milks to increase the overall acceptability of the products, as they can otherwise taste a little unusual to many of our palates. It's always worth checking the label as added sugar and a high content of oil (vegetable) means that some plant milks actually have a higher calorie count than whole milk. Added sugar in many plant milks is a clear difference from dairy milk but consumers can choose unsweetened versions if they wish. However, the higher salt content of plant milks may be a concern for some.

Plant milks are also generally much lower in fat than dairy milk. For example, rice and oat milks are lower in fat than semi-skimmed cow milk, whereas the fat content of hemp milk is higher (but not quite as high as whole milk). Other than coconut milk, the fats present in plant milks are mostly unsaturated. Coconut milk contains quite a bit of fat, in comparison with other plant milks, and almost all of it is saturated. In cow milk, around 60 to 65 per cent of the fat is saturated. None of the plant milks contains cholesterol. While there is some controversy about the role of saturated fats and their link to heart health, the advice remains that we should try and reduce our intake of saturated fats and replace them with unsaturated fats.

Pea milk and soya milk contain higher amounts of protein than other plant milks. Pea milk is promoted as having about the same amount of protein as dairy milk and soya milk is not far off this. While these milks may contain a good amount of protein, researchers point out that it is not equivalent to cow milk protein as cow milk contains more essential amino acids and has a higher DIAAS,[10] a measure of protein quality. Most other plant milks are very low in protein. The lack of protein, in particular, means that plant milks are not a great substitute for dairy milk for children. Children need more protein, as well as vitamins and minerals, to support their growth and development and dairy milk provides a good source of these nutrients.

10 Digestible indispensable amino acid score.

Unlike dairy milk, some nut milks contain a useful amount of vitamin E but, despite this bonus, plant milks typically require forti-fication with vitamins and minerals, like calcium, when used as a dairy milk substitute. Looking across brands, many plant milks these days appear to have been fortified with calcium and some vita-mins, particularly vitamin D and B vitamins, and many of these contain the same amount or more of these minerals and vitamins than dairy milk. However, the bioavailability of naturally occurring vitamins and minerals differs between dairy and plant milks, as well as those that are added. For example, it has been found that at least 30 per cent of the calcium in cow milk is absorbed by the body in comparison with 20 to 30 per cent from plant sources, such as almonds and beans. The type of calcium, or indeed vitamin, used to fortify drinks makes a difference to how much is actually absorbed, as does the type of drink it is added to, so you can't assume that the amount of calcium or vitamin D, for example, listed on a label is equivalent across products and they are not a guarantee of how much you'll absorb.

Calcium and vitamin fortification is helpful to improve the health value of plant milks but doesn't account 'for the full spectrum of nutrients available in dairy milk. One example is iodine. Dairy milk provides a good source of iodine but plant milks do not. A study of forty-seven plant-based milk alternatives, by the University of Surrey, found that overall they provided just a small fraction of the amount that is in cow milk and that just three of the forty-seven drinks were fortified with iodine. Many nutrition experts conclude that soya milk has the best overall nutritional worth as a replacement for dairy milk but it's not entirely comparable, and the taste and allergic potential means that it won't suit everyone.

To sum up, nutritionally, plant milks in general are not compara-ble or equivalent to dairy milk as their composition is typically quite different and not as nutrient rich. On balance, plant milks cannot generally be considered more 'healthy' than cow milk, because most contain less calcium, minerals and vitamins, less (and lower quality), protein, and more salt and often sugar. They do, however, offer a useful alternative to dairy milk for those who cannot have dairy for whatever reason. As dairy products make a significant contribution to the average diet and provide a good source of many nutrients, if you use plant-based alternatives it is important to compensate for the missing nutrients elsewhere in your diet.

So we've covered why plant milks don't quite match up to the marketing claims about their 'goodness', but could they be linked to health harms?

A key reason for a lot of people to turn to plant milks is to avoid the effects of lactose. Lactose is only found in dairy milk so all plant milks are lactose free and it is a no brainer to switch for those who suffer significant discomfort from regular milk. Plant milks also offer an obvious alternative for those who suffer from milk allergy. However, plant milks aren't the answer for all those with allergic tendencies and intolerances. Clearly, nut milks are not suitable for people with a nut allergy. Similarly, allergies to soya and sesame are well known and some oat milk contains gluten. In fact, many plant milks have been found to induce allergic responses so, for certain individuals, the choice may not be straightforward.

One of the main reasons that soya milk has declined in popularity over the years is fear about its safety, having been linked to breast cancer, amongst other illnesses. Soya has divided opinion and there have been a slew of studies investigating its effects on health. One aspect causing uncertainty is the high concentration of isoflavones in soya. Isoflavones are a type of plant oestrogen, similar in function to human oestrogen but with much milder effects. Isoflavones have complex interactions in the body and the findings of scientific studies vary depending on many factors, such as the ethnicity of participants, the type of soya product consumed, the intrinsic hormone levels of individuals and the part of the body affected. In terms of breast cancer risk, the World Cancer Research Fund assessed the evidence in 2017, and found no association between soya intake and an increased risk of breast cancer. Further to this, there isn't really hard evidence to demonstrate harms of regular consumption of soya milk in healthy people.

In the last couple of years you may have heard reports about the significant level of arsenic in rice. Arsenic is a natural metal element present in the environment and is absorbed by some crops as they grow. It is a toxic substance linked to significant adverse health effects, affecting multiple organs. It is known to accumulate in rice at higher concentrations than other crops (up to ten times the amount of other grains). But should this be cause for concern in rice milk? The arsenic content varies across types of rice and how it is cooked but the maximum concentrations have been found in rice bran, so products made from this contain more arsenic than other rice products. Rice milk is one such rice bran product and, because the arsenic is naturally occurring, there's no difference in arsenic content between organic and conventionally produced versions. One study of nineteen different rice milks found that all of them contained more than the EU limits set for arsenic in drinking water,[11] with up to three times this amount recorded for some. Before you panic, it seems that the average European diet is not actually that high in rice or rice products and typical arsenic exposure from these, and other sources, is within safe levels. But you probably shouldn't overdo it. Safety experts from Europe and the US agree that arsenic exposure from consuming rice products a few times a week probably does not pose a significant health risk but advise people to vary their diet by consuming a balance of different grains. Different countries have their own guidance regarding the consumption of particular rice products, and in the UK the Food Standards Agency advises that young children (under five years old) are not given rice milk as a dairy substitute due to the arsenic content. Devotees of rice milk

11 Whether rice milk is categorised in the same way as water is another matter.

might like to consider incorporating other plant milks into their regular diet and alternating between those and rice milk.

Concerns have been raised (by which I mean social media panic) about the inclusion of carrageenan in foods, including nut milks. Carrageenan is a widely used thickener that is extracted from seaweed. A small number of researchers in the 1990s linked it to gastrointestinal diseases and other health complaints. As products like nut milk are consumed as part of the daily diet, the regular consumption of carrageenan worries some people. However, many more scientists have since shown the research to be flawed and have repeatedly demonstrated that there are no safety issues regarding carrageenan. Bodies, such as the US Department of Agriculture, the Food and Agriculture Organisation and the World Health Organisation, are clear that carrageenan is a safe product.

Environmental impact

Some consumers have been turning to almond milk over concerns about the environmental impact of eating animal products, but they may not be aware of the significant environmental impact that almond cultivation is also having. Although almond milk production produces fewer greenhouse gases than cow milk production, it requires significant water usage and pesticide use. To put that in context, did you know that it requires around 74 litres of water to produce a single glass of almond milk? Some researchers put this estimate even higher. Cow milk production uses even more water and produces 1.31kg CO_2e[12] more per litre than almond milk, however. If you're worried about the impact almond milk may be having on the environment be aware that it is not the only plant milk to have a negative image in this regard.

Soya has also got a bad reputation for its effect on the

12 Carbon dioxide equivalent (CO_2e) is a term used to describe different greenhouse gases in a common unit.

environment. Together, the US, Brazil and Argentina produce about 80 per cent of the world's soya. Large swathes of land are being deforested (including rainforests) to make way for the increasing demand for this crop. According to the World Wildlife Fund, soya is the second largest agricultural driver, after beef production, behind deforestation worldwide. This is clearly a big issue. The UK imports about 75 per cent of its soya beans from the Brazil and the US. Of course, you can't pile all the blame on soya milk consumption. In fact, a vast proportion of soya is used as animal feed, as well as in a huge range of human food products and even biodiesel. Efforts are being made by the soya industry to address the problems, and some supermarkets are trying to stock only soya products from sustainable sources, but there's still some way to go.

A lot of soya harvested for soya milk products is genetically modified (GM). As much as 90 per cent of the soya produced by the US, Brazil and Argentina is GM. The discussion about GM foods is for another time but regardless of the evidence surrounding it, many consumers are uncomfortable about the prospect of GM products and are often keen to avoid them. This has also played a part in the declining sales of soya milk. It is worth noting, however, that not all soya is genetically modified and you can often check for this on the product label if it is something that worries you.

There will likely be some environmental impact of all plant milks but some more so than others. It's worth reminding ourselves that for those of us in the UK, the key ingredients of most plant milks are cultivated in countries far away so their carbon footprint from travel alone is substantially higher than dairy milk produced in the UK.

Choosing your plant milk
Given the general lack of unique nutritional worth of many plant milks, some of the possible concerns about some of them, and the fact that they are mostly water, for those who have the choice it may

even raise the question of why drink them at all. Now you know what is in commercially made versions of plant milks, you might prefer to make your own at home as they're pretty simple, but bear in mind that you'll need to make up some of the nutrients from elsewhere in your diet. Choosing your plant milk depends very much upon your motivation as they offer different selling points. For instance, you may have particular allergies, which will narrow down the options, or you may be looking for a high protein option, or want something low calorie or something that doesn't wreck the environment. You may simply just want the one that tastes nicest. There are now so many choices that it's probably worth having a think about what you really want, or need, from your plant-based drink/milk alternative/mylk.

As adults, many of us would likely squirm at the thought of drinking human breastmilk but don't think twice about drinking the 'breastmilk' of another species. When put like that, perhaps drinking animal milk is just a bit, well . . . weird? But the thing is, regardless of the ethics of drinking it, animal milk is nutritionally rich and provides a quick way to gain many essential nutrients at once, whereas dairy milk alternatives don't really compare, gram for gram. In fact, it would be better to consider them as completely different foods, then you would be more likely to balance your diet around them accordingly. While many of us are lucky enough to be able to debate the pros and cons of different milks, it is worth remembering that animal milk plays a vital role in the diets of children in populations with very low fat intakes and limited access to other animal source foods. It would be highly challenging to produce human breastmilk for adult consumption on a wide scale (and I'm not sure many lactating women would be up for that proposition) so the milk of other animals is probably the next best thing, nutritionally speaking. It's not for everyone, though, so plant-based milks provide a range of convenient alternatives, provided they are part of a broader, well-balanced diet.

World's most expensive milkshake[13]

Back to those world records . . . The record was set for the most expensive milkshake at US $100 in June 2018. So what do you get for your money? The LUXE milkshake combines Jersey cow milk, known for its high butterfat content, with Tahitian vanilla ice cream, Devonshire clotted cream, Madagascar vanilla beans, 23-carat edible gold, whipped cream, rare donkey caramel sauce and gourmet maraschino cherries. And it is all served up in a specially designed glass adorned with over three thousand Swarovski crystals. If you want one, you'll need to head to the *Serendipity 3* restaurant in New York where it was created.

---

13 A milkshake is a sweet, cold drink that combines cold milk with additional ingredients, such as ice cream and flavourings. It is shaken or blended to mix everything thoroughly and the result is typically a thickened, slightly frothy drink.

CHAPTER THREE

Hot Drinks

....

How well do you know your cha from your chai? What about a decaf, skinny, soy mochaccino? When I was child, you ordered either a 'tea' or a 'coffee' in a café – there was no discussion about what type or how it was to be brewed and served. Milk was typically added and the only optional extra discussed was sugar. Boy, how things have changed since then?! Nowadays, we spend many hours of our lives standing in slow queues waiting for endless orders to be discussed, deliberated and then slowly churned out by ever-more complex machines behind the coffee shop counter. Some of us tense up in a panic because we don't understand the options we've just been asked to choose from by the barista. Whoever predicted that advances in technology would speed up business production over the years had clearly not taken hot drinks into account. Ordering a hot drink these days is a time-consuming, complex and, often, stress-inducing operation.

In this section, I'll focus on the most popular hot drinks consumed in the world, namely tea, coffee and those that are cocoa-based. This focus is partly due to their global popularity and partly due to having had more health research carried out on them than other hot drinks. Of course, there are many other hot drinks that are also popular that I don't have space to do justice. For example, atole is a popular thick Mexican winter drink, a warming and comforting blend of cornmeal and water, mixed with a range of different ingredients. It may be served sweet or savoury and added ingredients include sugar, honey, fruit, chocolate, cinnamon, nuts and chilli. Bouillon drinks are preferred by some, not forgetting Bovril, the beef extract particularly favoured in the nineteenth and twentieth centuries that many of us remember from our child-hoods. There's also the simple soothing mug of warm milk or hot blackcurrant squash, and an array of alcoholic hot drinks not covered here. But that's enough of what I'm *not* talking about: let's get on with what I *am* talking about.

Hot drinks: in essence

In 1992, Stella Liebeck bought a take-away cup of coffee from a McDonald's drive-thru in Albuquerque, US, not realising that what was about to unfold would hit the headlines around the world and continue to be talked about today. While sitting in a car, seventy-nine-year-old Stella spilled the coffee into her lap causing serious burns to 16 per cent of her body, leading to hospitalisation and years of skin grafts and other treatment. The injuries from the scalding coffee were caused in just three seconds and the life-changing nature of these events prompted a lawsuit. Stella Liebeck sued McDonald's for damages, to cover her medical expenses and lost income, and she won. The Liebeck versus McDonald's lawsuit became famous around the world as it was portrayed as a combination of America's over-litigious culture with a presumed carelessness by one individual. We've all spilt drinks in our time so the case felt like this individual was attempting to make some money out of their own clumsiness. However, when looking at the facts more closely, the whole case was actually about corporate risk-taking and recklessness. It turned out that the coffee served was dangerously hot. McDonald's required employees to serve coffee at a temperature of 180 to 190°F, over 30 to 40 degrees hotter than many other companies and home coffee machines, that experts testified was an unacceptably high level of risk. Furthermore, there had been over seven hundred previous injury claims brought against McDonald's due to coffee burns and the company admitted to knowing for at least ten years about the risk of serious burns from its coffee. Despite the case, the company still serves its coffee very hot and further burns cases have been made against it. So why do they serve such hot coffee? At first they said it was to allow it to cool to the right temperature by the time people got to work or home, despite their own research showing that most customers drank the coffee while still in their car. They later claimed that a consultant advised that this temperature range was the best for taste, but counter to this McDonald's own quality assurance manager testified that their coffee was not fit for consumption at the temperature at which it was poured into cups as it would burn the mouth and throat. Whatever the reason for the very hot coffee, the debate continues to percolate.

Whilst writing this section of the book I was unwell for six weeks or so and during that time the only drink I could face was water; coffee felt too rich and tea too acidic. Once my tastes just about normalised again, I decided to have a cup of rooibos like my old routine. I can only describe the experience of that first new cup as totally alien. The sensation of heat passing through my throat and mouth felt really strange, and not in a good way. It was unpleasant and it took me a few more weeks before I properly got back into drinking hot drinks again. It made me wonder whether drinking hot drinks is something we slowly develop a taste for rather than have an intrinsic desire for.

Before throwing the spotlight on individual drinks, let's get under the skin of caffeine, a major component of popular hot drinks.

What is caffeine?

Caffeine is something that many of us feel we simply couldn't function without. It is a naturally occurring stimulant, from the methylxanthine class of drugs. Methylxanthines are found in high concentrations in tea, coffee and chocolate. Caffeine stimulates the central nervous system, increasing alertness and potentially inducing agitation. Other than tea, coffee and chocolate, caffeine is found in colas, energy drinks and other botanical sources, like guarana, kola nuts and yerba mate. It is also used in a range of medications, such as painkillers and migraine treatments. As well as helping medicines be more effective and absorb into the body more quickly, caffeine has anti-inflammatory properties.

> **Read the label**
>
> You may not realise it but many other (some admittedly niche) products we use or consume contain caffeine, such as lip balm, body lotion, shampoo, chewing gum, waffles and beef jerky.

This table below gives you an idea of how much caffeine is in some of our most popular beverages. (This is just a rough guide, however, as you will discover further on that different types of tea and coffee, for example, vary considerably in their caffeine levels.) In case you're wondering, for comparison, a 50mg bar of dark chocolate contains less than 25mg of caffeine and the same size milk chocolate bar contains less than 10mg.

	Approximate caffeine content per serving (in milligrams)
Mug of filter coffee	140mg
Mug of instant coffee	100mg
Mug of decaffeinated filter coffee	2–8mg
Cup of tea	50–75mg
Cup of decaffeinated tea	2–5mg
Cup of green tea	40mg
Coca-Cola*	32mg
Coke-Zero*	32mg
Sprite*	0mg
Fanta Orange*	0mg
Relentless energy drink (250ml can)**	80mg
Red Bull energy drink (250ml can)**	80mg
* Based on UK standard can size – 330ml	
** Also comes in larger can sizes, increasing the caffeine content	

There aren't accurate statistics on caffeine consumption world-wide although a number of studies provide estimates for a few countries, some based on diet surveys, others based on coffee sales. From a range of studies in recent years, it's been estimated that: a) over 85 per cent of adults in the US consume caffeine daily, with an average intake of 180mg per day; b) in the UK, on average, adults consume around 130mg of caffeine per day; and c) in Australia the intake is somewhere between US and UK values. None of these statistics is totally up to date, nor gathered by the same method, but they give us a rough idea of what we're probably consuming on a daily basis, although our individual intakes will vary widely. Estimates from the EFSA show an average range of caffeine intake for eighteen- to sixty-five-year-olds as being between 37 and 319mg per day.

One large review of caffeine consumption around the world found that, despite the introduction of many new caffeinated foods and drinks in recent years, total caffeine intake has remained stable over the last ten to fifteen years. Coffee, tea and soft drinks continue to comprise the majority of caffeine consumed and, perhaps surprisingly given the amount of media attention they receive, energy drinks apparently contribute very little to our total caffeine intake.

In most European countries, coffee is the predominant source of caffeine but in Ireland and the UK it is, predictably, tea. However, this may not always be the case. Back in 1975, we in the UK purchased, on average, 66g of tea per person per week. Fast forward to 2015 and that figure had dropped to 24g. While tea purchases have seen a drastic decline over the decades, coffee purchases have steadily risen and, if these trends continue, it probably won't be long before they match tea sales. Our long-held reputation as a nation of tea drinkers may well be at risk.

How are decaffeinated hot drinks produced?
I gave up drinking caffeine around fifteen years ago, during a time when I had really poor sleep and was trying anything in an attempt to improve it. Tea has always been a life source in my family and the kettle is rarely quiet, so the transition was tough. For a while I convinced myself that my decaf self must be somehow healthier than before, but now I wonder whether I should be drinking it after

all, given that we're led to believe that some caffeine is, in fact, good for our health. So, is this really true and should I consider falling off the decaf wagon?

There are many people, like myself, who want to avoid regular caffeinated drinks for any number of reasons, but still enjoy a nice cuppa. These days we're fortunate to have a wide choice of decaffeinated alternatives available. A slight caveat to this is that you'll have noticed from our table of drinks that decaf tea and coffee is not entirely caffeine-free and still contains a small amount. But if caffeine is a natural component of tea and coffee, just how is it (mostly) removed?

There are three main methods that are used to remove caffeine from coffee. They share the main stages of the process but differ in their extraction method. First, green (unroasted) coffee beans are swelled with water or steam to make the caffeine soluble so it can be removed. Extraction then takes place. After this, steam stripping removes any remaining solvents used in the extraction process, then the beans are dried back to their normal moisture content. Let's focus on the different extraction methods.

The water method (often called the Swiss Water Process) immerses the coffee beans in water. Caffeine is water soluble so over time the caffeine is drawn out and then removed when the solution is passed through a carbon filter. The remaining solution is returned to the beans for reabsorption of flavours and oils that were also drawn out. The downside of this method is that some of the coffee's aromatic quality might be impaired. To counter this, liquid coffee extract (from which the caffeine has already been reduced) is added to the water that is circulated around the beans. Although it seems to be the most appealing method from the perspective of the consumer, the water method is not the most efficient at removing caffeine and not very selective, in that it removes other desirable qualities of the coffee. This is why some manufacturers employ other methods.

The majority of decaffeinated coffees are produced using solvents. The solvent method uses one of two approaches: direct and indirect. The direct approach circulates solvents, such as ethyl acetate or methylene chloride, through a bed of coffee beans. Such solvents are efficient at removing caffeine and the solvent plus caffeine is evaporated off. The beans are then washed with water. The indirect approach first soaks the beans in water to remove the caffeine (remember that it also draws out other flavours and oils). The water containing the caffeine (and other flavours etc) is then treated with a solvent to remove the caffeine, and heated to get rid of said solvent that has captured the caffeine. The remaining liquid, with all its flavour goodness, is returned to the beans to be reabsorbed. The process is repeated a number of times to reduce the caffeine content to the desired level. This indirect method is so-called as the solvent never directly comes into contact with the coffee beans themselves.

The third method uses carbon dioxide (CO_2). Liquid CO_2 is forced into coffee beans at high pressure and acts as a solvent to draw out the caffeine. The caffeine-filled CO_2 is then returned to a gaseous state and the caffeine removed. CO_2 is highly selective in just removing the caffeine and no other desirable components but is an expensive process and unlikely to be used on small batches of gourmet coffee.

Many coffee lovers would rather drink something else entirely than be subjected to a decaf coffee. They just don't see the point of it. One complaint about decaf coffee is that it is notoriously challenging to find a good one. There may be a good reason for this. We've seen how the decaffeination process itself may reduce some of the flavour, although efforts are made to minimise this, but the roasting step may also be to blame. Interestingly, decaf coffee is particularly difficult to roast. During the roasting process, coffee beans change in colour and composition. The longer they're roasted, the darker they become and the flavour profile changes accordingly. The problem with beans that have had their caffeine removed is that they darken during the decaffeination process and in appearance look more developed than they actually are. They are also lighter in weight so respond differently to heat. This makes it difficult to judge how long to roast them for – the colour may need to become darker than regular beans during roasting because they were darker to start off with, but they may roast more quickly due to their light weight. Getting the balance right is tricky, hence the variable quality of decaf coffees on the market.

The methods for removing the caffeine from tea are essentially the same as the options for decaffeinating coffee. The exception is that the water method is not typically used as soaking the tea leaves in water is considered to water down the flavour (literally).

Some of the treatments used in decaffeination processes have caused alarm as to whether they might be of harm to our health, but let's look at the evidence for this. Although it may sound somewhat industrial, ethyl acetate (EA) is naturally found in fruits and vegetables, including coffee, and in alcoholic drinks. A synthetic form is used in the decaffeinating process. Ethyl acetate may cause ill-health effects if inhaled but has not been found to cause problems when ingested. While methylene chloride (MC) is used as a solvent in food technology, it is also used in paint removers, aerosol formulations, and in the manufacturing of pharmaceuticals and electronics. Again, inhalation of this solvent is known to cause harm, but there is also some debate as to whether it might also lead to problems when ingested. There is evidence to suggest the MC is a probable carcinogen (i.e. has the potential to cause cancer), but much of this data is from animal studies and based on inhalation of significant amounts rather than ingestion. Food regulators have deemed it safe for use in decaffeination. Having said all this, the amount of solvent likely to remain after the decaffeination process is miniscule as any residues are unlikely to survive the steaming, roasting and brewing processes. This is because they are volatile, meaning that they easily evaporate at relatively low temperatures. EA has a boiling point of 77°C and for MC this is lower, at 39.7°C. The steaming process applies high heat to evaporate off solvents. If there is any solvent left, it will be subjected to further high temperatures as part of normal coffee and tea production. Coffee beans are roasted from anywhere between 180 and 240°C. In addition, both tea and coffee are brewed in hot water at temperatures of 70°C and above. So, any tiny traces would likely evaporate before you drink your hot beverage.

But can we reduce our caffeine intake and still benefit from other advantageous components of tea and coffee? There is confusing and contradictory information about whether decaffeination processes remove significant amounts of other components of coffee and tea, such as those that are implicated in health effects (e.g. antioxidants). Despite this, it seems that, on the whole, many such features are still usually present in decent amounts and so decaf tea or coffee may affect our bodies in similar ways to their full-caf cousins. I'll talk more about this later.

Caffeine and health

Caffeine benefits (some) babies

Caffeine is frequently given to premature babies to stimulate lung development and help their brains to remember to breathe. The babies are weaned off caffeine once their lungs have developed enough to breathe on their own.

Many potential benefits of caffeine have been mooted and numerous studies have looked, and are still looking, into this. I'll delve into this later as most of the studies investigate the effects of specific beverages, such as coffee or tea, rather than caffeine itself.

Despite potential health benefits of caffeinated drinks, there's no official advice recommending we consume a certain amount every day. However, caffeine is also known to have harmful side-effects so limits have been established to guide us on how much might be too much of a good thing. Too much caffeine can lead to restlessness, increased anxiety, raised heart rate, headaches and nausea, as well as behavioural problems in children. Caffeine in high amounts has also been linked to bone loss. How much is too much can depend on the individual and their sensitivity to caffeine. Most adults don't need to worry too much about their caffeine consumption as, on average, our intake is within recommended limits (up to 400mg per day for adults). But, as well as people with high blood pressure, there are some groups who should restrict the amount they consume.

Pregnant women are advised to limit their intake to 200mg per day or less. The reason for this is that high levels of caffeine during pregnancy have been found to result in a low birth weight, which in turn may lead to health problems for the baby, and very high caffeine intakes can increase the risk of miscarriage. Children and adolescents are thought to be particularly sensitive to the effects of caffeine as their brains are still developing. They may also be more affected by caffeine's stimulatory properties due to their smaller body size. In most countries, chocolate, tea and soft drinks make up the main sources of caffeine in children. Another group who need to be a little careful about caffeine are some people with mental ill-health. If you're already suffering anxiety or insomnia, for example, caffeine may exacerbate symptoms.

Although the stimulatory effects of caffeine are rapid, with effects being felt within 15 to 30 minutes, it hangs around in the body for much longer, and can be present for eight hours or more. This is the reason why some people are advised not to consume any caffeine after lunchtime, to avoid it interfering with their sleep at night.

I've talked about a range of effects of caffeine in general but can we pick out whether tea, coffee or other caffeinated drinks are having specific effects?

A nice cup of tea

> ### Who drinks the most tea?
>
> In 2016, the people of Turkey were the world's biggest tea drink-ers, consuming nearly 7lbs of the stuff per person. The Republic of Ireland and the UK are the next largest consumers, per head of population. Although China is the largest consumer in total, due to the size of its population, its annual consumption per person doesn't even make the top ten list.

The tea plantations and the Dodabetta Tea Factory in Ooty, high in the hills of the Tamil Nadu region of India, proved an unforgettable experience when we visited some years ago. The freshness of the tea plants in the clear air of the hills, as well as the aromas released during the tea processing linger in the memory. The intricacy of how the tea we drink is produced gave me a finer appreciation for this traditional beverage.

Tea, cha, brew, cuppa, infusion, tisane . . . all words describing a hot drink usually based on some sort of leaves. True tea is a hot beverage made by infusing the dried and crushed leaves of the tea plant in boiling water. Humans can't get enough of it and it is the world's second most widely consumed drink, after water, cherished for its stimulant properties and health benefits.

There are many varieties of tea, each with unique properties, and these are typically grouped into five main categories: white tea, green tea, oolong tea, black tea and Pu-erh tea. (There is also yellow tea, a slightly fermented Chinese tea, but it is a rare commodity.) By far the most widely drunk tea in countries like the UK and US is black tea, whereas green tea is more popular in East Asia. All teas derive from the leaves of the same plant, *Camellia sinensis*, a sub-tropical, ever-green bush native to Asia. There are two recognised varieties: *Camellia sinensis* var. *sinensis* (Chinese tea) and *Camellia sinensis* var. *assamica* (Assam tea or Indian tea). The tea plant needs a hot, moist climate to thrive, prefers acidic soils and ideally a slope of 0.5 to 10 degrees at elevations up to 2000m, and so is geographically limited to a few areas around the world. China and India are the world's two largest tea-producing countries.

So if all tea comes from essentially the same plant, what defines the different categories? Well, it all comes down to how the tea leaves are processed.

After the leaves are plucked from the plant, they are taken to be processed. The leaves are laid out to dry, in a process called 'withering', so they become supple for the next stage: rolling. Tea leaves are rolled and shaped, which breaks down the cell walls, releasing enzymes and oils that will alter the leaf's flavour and exposing them to oxygen, initiating the oxidation process. The oxidation process determines the colour, taste and strength of the tea and involves the leaves being exposed to oxygen for a certain period of time. The longer the leaves are left to oxidise, the darker in colour and stronger in flavour they will become. So, as you might imagine, black teas are highly oxidised whereas white and green teas are the least oxidised. Once the desired oxidation level is reached, the leaves are fired to halt the oxidation process and further reduce the moisture content to ensure the tea keeps well. After processing, tea leaves are sorted and graded by leaf size and colour to create different batches of tea.

Not all teas go through all the stages of the process, and some may go through different stages a number of times. This is how leaves from essentially the same plant become differentiated. White tea undergoes the least processing and produces the most delicate of teas. It is made from the newest buds on the plant and after withering moves straight to the drying stage. It is neither rolled, shaped nor allowed to oxidise. Green tea is withered, rolled and dried, avoiding the oxidation process. A light, fresh flavour results from the lack of oxidation. The leaves are often steamed to help stop any oxidation that might occur from the rolling stage and this helps bring out the fresh flavour. Black tea goes through the entire process and is allowed to oxidise completely. Oolong tea falls somewhere in the middle. Although it follows all the stages, it isn't allowed to fully oxidise. The amount of oxidation ranges widely and, as such, this category of tea varies the most in its flavours and aromas. Pu-erh tea is different again. The process is similar to that of green tea but before the leaves are dried, they are pressed into shapes (or tea 'bricks' or 'cakes') and aged. The flavour profile can change considerably over time. Like fine wines, some Pu-erh teas are many years old and there are connoisseurs who enjoy collecting such teas, prized for their earthy, rich tastes.

What I've described above is known as the orthodox process. There is also the non-orthodox process, called the crush-tear-curl (CTC) method that originated in the Second World War to increase the amount of tea that could be packed into a tea chest. In this process, after the tea leaves are withered, they are put into a machine which crushes them, tears them and curls them. The crushed leaves are formed into pellets and then oxidised and dried. This process is much quicker than the orthodox method and the resulting pellets are ideal for teabags. The CTC process is most often used to produce black tea. You can probably deduce that CTC tea tends to be lower in quality than tea produced by the orthodox method and is more bitter and less diverse in its flavour profile.

Preparing the perfect cuppa

So, you've got your tea leaves (possibly in bags) but what should you do next to get the best out of them? Going by what the experts advise, it seems that most of us are probably making our tea all wrong.

First, the water you use affects the tea. Softer water results in a cleaner finish as the minerals in hard water can result in a scummy layer on the tea. Some say it's best to use the water you're used to and make sure it is fresh. Fresher water seems to bring out a brighter, cleaner taste. Next, we need to think about temperature. We tend to stick the kettle on and when it boils pour the water on our chosen tea, but this might not be the best method for all teas. Different teas require different brewing temperatures to make the perfect cup. How long you steep, or brew, your tea for will also affect its flavour: too little time and the result may be rather weak and watery, too long and the tea may become bitter. So what's going on here?

While there are a various guides on how hot individual teas like their water, the following pretty much sums it up: darker, stronger teas, such as Pu-erh, black and oolong teas, need a higher temperature and longer time to brew, whereas milder, more delicate teas (i.e. green and white) require milder temperatures and shorter brewing times. However, most teas favour their water under boiling temperature (take note, British readers!) and brewing times range from around two to five minutes. It's a matter of personal taste, of course, but if you prefer strong tea, it is better to add more leaves to enhance the flavour rather than leave it to brew longer, which will just increase its bitterness.

As tea brews, the leaves release tannins, amino acids, aromas and flavours. These compounds on the surface and inside the leaves slowly diffuse into the water and this process will vary in the time it takes depending on the compound, the tea type and water temperature. Aromas responsible for smell and flavour dissolve pretty instantly into the water (although compounds related to mouthfeel and texture take longer), lighter polyphenols and caffeine take a little more time but still dissolve fairly quickly, and the heaviest compounds (including heavier polyphenols, flavanols and tannins) take the longest time to pass into the water. The aim when brewing tea is to bring out the best flavours and nutrients, while avoiding bitterness (caused by the tannins). The longer tea steeps, the more tannin is released (more on tannins in a moment). During brewing, a peak is inevitably reached when the tea is at its best and after which the tea becomes bitter and unpleasant. Depending on the particular tea, too much heat may dissolve the tannins and flavour compounds too quickly creating an imbalance, and too little heat may have the opposite effect, resulting in a weak, flavourless brew.

It's not just the tea type and water you need to get right. Your choice of teapot also matters, and can affect the time and quality of the brewing process. Teas that require a higher temperature to brew perfectly benefit from a teapot that will retain the heat as it brews, like one made from metal or ceramic. Conversely, teas that prefer a lower temperature need a teapot that better releases the heat over time, such as a glass or porcelain one. And don't forget to warm the pot. This helps to create the right temperature for brewing. The same principle goes for the cup too. Certain materials help your tea retain its temperature for longer but also mean it takes longer to cool to the best drinking temperature (which is around 60°C, if you're wondering; above this, and you're forced to take in extra air to cool it or, in other words, slurp your tea). A friend asked me to find out why tea from polystyrene cups always tastes awful. A number of theories have been suggested: 1) that the tea stays too hot and you can't drink it properly; 2) that you prejudge the tea as being inferior before you taste it because it comes in a polystyrene cup; 3) that somehow the polystyrene is transferring some of its own taste to the tea; and 4) that the cup is absorbing flavour molecules from the tea. Unsurprisingly, none of these have robust evidence underpinning them. It's probably worth bearing in mind that when you have tea served in a polystyrene cup, you're unlikely to be in the frame of mind to savour and enjoy it – think about when you've had tea in such a cup: In a work meeting? At a conference? Pacing a hospital corridor? Trying to keep warm at an outdoor festival? None of these situations is what you might consider relaxing or focused on the tea.

The microwave: a tea-making *faux pas*?

Hands up who uses the microwave to make tea . . . I know a few people whose hand would be sheepishly raised at this point. The convenience is appealing but our gut feeling suggests that using a microwave can't be a great way of preparing tea. Does it really make a difference to the quality? Luckily for us, scientists have actually spent time looking into this. It turns out that on the one hand, as water in a microwave is heated unevenly and its temperature is difficult to control, it may not be the right temperature to properly brew the tea. On the other hand, a study by a team in Australia found that employing the microwave to make green tea was more effective than traditional brewing for extracting caffeine and beneficial compounds. Their method involved putting a teabag in a cup, adding boiled water, leaving it for 30 seconds and following this with 60 seconds in the microwave. Marvellous stuff. But the scientists point out that this method also produces tea that is rather strong in flavour and may be more bitter than recommended by the tea manufacturers. I'm not sure it's really worth all the palaver.

Now let's talk about milk. Many of us enjoy milk in our tea. It hails back to a time when common china cups weren't as strong as they are now and would shatter when boiling liquid was poured in. They literally couldn't stand the heat. So, milk was poured in first to offer some protection against the hot tea. High quality bone china of the day didn't have this problem, however. As milk was widely added to tea, whether it was poured in before or after the addition of tea was an indication of social status: if you put the milk in first, it indicated that you couldn't afford the finest bone china, whereas if you added the milk after you were showing that you had china that could withstand the tea's heat and, basically, flaunting your position in society. We've been adding milk for centuries and it's a tradition that has stuck, but does it make any difference to the tea itself? There's some research to show that milk binds with the polyphenols in the tea, reducing their bioavailability and thereby inhibiting beneficial antioxidant effects. However, the evidence is conflicted as others have shown that the same amount of antioxidants reaches the bloodstream regardless of whether milk is added or not. So, it is not yet clear whether the addition of milk makes a significant difference to the health benefits of tea. In terms of taste, Dr Andrew Stapley, a chemical engineer from Loughborough University, claims that science proves you should add the milk to the cup before the tea and not after. His experiments showed that if you add the milk after the tea, proteins in it clump together as they encounter the very hot temperatures but apparently this is less likely to happen if the milk is put in the cup first. And the clumping proteins means the tea doesn't taste as nice (although this is up for debate). If you're interested, the Royal Society of Chemistry has developed their own guide to brewing the perfect cup of tea, based on science. Or you could trust your own taste buds, just as humans have been doing for hundreds of years.

To bag or not to bag?
That's all very well, you might think, but what about plain old teabags? Can we still just stick them in a mug and whack on the boiling water? You've probably guessed that much of what I've been referring to until now is relevant to loose-leaf tea but most of us actually use teabags. Teabags have been around since the early twentieth century and were willingly adopted by the US

population. The same could not be said for people in the UK who were rather skeptical at first. It wasn't until the 1950s, and the booming trend for labour-saving devices, that teabags were finally accepted in this part of the world. At the start of the 1960s teabags made up only 3 per cent of the UK tea market, but that number grew quickly and today it is estimated that 96 per cent of UK tea drinkers use teabags.

Teabags are mostly made from paper of vegetable and wood fibres but many of the most common varieties we use also contain plastic (polypropylene to be exact) that is used to seal the bags. Consumers have become increasingly unhappy about this, primarily on environmental grounds in that the bags don't readily degrade,[1] and some manufacturers are now changing to entirely plant-based alternatives (such as corn starch). There are also other bags, typically containing more expensive teas, that are made from nylon or polyethylene terephthalate mesh (think of the pyramid-shaped ones). These are obviously not biodegradable and questions have also been raised over whether heating the plastic in hot tea results in something less safe to drink. There's currently no good evidence to show such teabags pose any real risk to health but there haven't been large studies specifically investigating this. Many teabags are bleached white for aesthetic purposes – it's thought they look more appealing than the darker, naturally tinted bags. The bleaching process may involve chlorine, oxygen, ozone and/or hydrogen peroxide. This concerns some consumers and there in an increasing shift towards unbleached teabags as manufacturers recognise consumer demand for more natural products.

Commercial teabags are often considered inferior to loose-leaf tea. Higher quality leaves are used for loose-leaf tea whereas the

---

1 You might think that a few teabags failing to rot down is not that much of an environmental impact but when you consider how many billions of teabags are produced and used around the world it builds up to a much more significant issue.

vast majority of teabags contain low-grade leaves that have been broken into small pieces (called fannings and dust) by the CTC method. As loose tea steeps, the leaves infuse and expand, allowing the water to flow around them, gently releasing their natural compounds. However, the ability of teabag tea to infuse is limited as it doesn't have as much space to expand. The processing method and confinement of the bag are thought to result in a less diverse flavour profile.

The CTC method often produces tea in high volume that can then be stored for a long time. The longer storage time means the tea inside the bags may not be as fresh as loose-leaf tea, further impairing its flavour potential. While there aren't radical differences in the levels of antioxidants found in teabag tea versus loose-leaf tea (studies have found more in one, or more in the other, or little difference), longer storage times do reduce antioxidant levels and this is perhaps more pertinent to tea in bags. Tea, particularly green tea, is also rich in compounds called catechins that are thought to contribute to the beneficial effects credited to this beverage. The greatest amount of catechins are found in fresh leaves and they degrade over time once picked so, inevitably, tea that has been stored longer will contain fewer catechins. This is exacerbated in teabag tea because as the leaf pieces are much smaller than whole leaves, they have a greater surface area that means they're more exposed to light and oxygen that degrade catechins. This also affects flavour as aromatic oils are more vulnerable to evaporation. (This is why tea that you've had hanging around in the back of the cupboard for ages just doesn't taste that great.) The larger surface area also means they can interact with water more efficiently during brewing, meaning a greater release of tannins, which is why optimum brewing times for teabag teas are typically shorter than loose-leaf teas. The upside to this is that they also better release constituents like L-theanine, which contributes to the relaxing properties of tea. If you've ever wondered why tea is considered relaxing whereas coffee is thought of as stimulating, but they both contain caffeine, then L-theanine is part of the answer. It is a unique amino acid found in tea but not coffee that helps us to relax without causing drowsiness.

When it comes to the handy teabag, one thing is for certain . . . you definitely don't want to add the milk first when brewing a teabag directly in the cup. The milk makes it too cool for the tea to brew properly, resulting in a weak (and, frankly, disgusting) cuppa.

The office tea run

Having a tough day at the office and looking forward to that morning tea break? You might want to think carefully about where your cup of tea is going to come from to avoid your day getting any worse. A swabbing study of offices in the UK found that the tea run may be a potential bacterial health hazard. On the tins or boxes where the tea was kept, the researchers recorded microbial levels seventeen times higher than you would find on a toilet seat. The kettle, fridge door handle and sugar bowl were also culprits and using someone else's mug was also found to raise the risk of colleague contamination. Poor hygiene, not washing hands or mugs before making tea, was blamed.

At the end of the day how you have your tea is a matter of personal taste (who am I to judge?) but if you want to see what your tea should taste like in optimum conditions, you could try following the observations above and perhaps you might become quite the connoisseur.

What's in your tea and is it as good for us as hyped?

People have been drinking tea for its medicinal properties for centuries but what exactly is in tea that's so beneficial? Tea is filled to the brim with interesting, and potentially beneficial, compounds. It contains thousands of chemical components, with over seven hundred aroma compounds alone. As well as those compounds, tea is full of polyphenols, amino acids, enzymes, methylxanthines, minerals and vitamins that demonstrate an array of bioactive properties. So much so, that many individual compounds of tea are being explored for their potential as functional food ingredients (i.e. ones that could be used on their own, or in combination, and often in higher doses, to fortify other foods or supplements to enhance their health benefits). While teasing out exactly which compounds are good for us, if any, is tricky, the three main bioactive compounds[2] in tea are considered to be catechins, caffeine and L-theanine.

Polyphenols

Polyphenols are the main compounds in tea and primarily affect its astringency[3] as well as some of the flavour and colour. There are an estimated thirty thousand polyphenolic compounds in tea. Polyphenols have been found to have antioxidant properties, meaning that they prevent or fight off damage in your cells, and some have shown anti-inflammatory and anti-cancer effects. In black tea, the most abundant polyphenols are tannins and in green tea, the most abundant polyphenols are catechins.

2 Compounds that are not essential nutrients but thought to impact health.
3 The puckering or drying sensation in the mouth due to the polyphenols' reaction with saliva. Polyphenols bind to saliva, leaving the mouth dry.

Tannins

Tannins (water-soluble polyphenols) are commonly found in plants and fruits. They act as a defence mechanism against attack from disease and animals, which is why they typically produce an unpleasant or bitter taste to induce a negative response in whoever is attacking the plant. In tea, while too high a concentration of tannins can produce a bitter taste, a moderate level also contributes to the richness of flavour, giving tea body. Tea lacking tannins would taste weak and watery. There is a diverse range of tannins, with differing properties. Tannins are sometimes referred to as 'antinutritional' as some research suggests that they can inhibit the effects of other nutrients. For example, in the past it has been well documented that tannins impair the bioavailability of iron in the diet and may, as a result, decrease a person's iron intake. However, the evidence is not consistent and a review of studies looking into the effect of tannin on dietary iron indicated that individual iron status is not often affected by overall tannin intake. Tannins bind to proteins, including the milk in your tea, which contains a protein called casein. This is the reason that, after drinking, a cup that contained tea with milk often looks cleaner and less stained than one that held black tea; the tannins responsible for the discolouration bind to the milk and are drunk along with the tea. Some recommend that adding milk to your tea helps to keep your teeth whiter (versus drinking tea without milk) as many of the tannins are bound up in the milk and not available for staining your enamel. Tannins have also been found to have some benefits, such as anti-cancer and antimicrobial properties.

Tannins theaflavin and thearubigin in black tea impart colour, astringency and body. Some research has found that they impart similar antioxidative properties as catechins in green tea and they have also been linked with health benefits, although more research is needed to delve sufficiently into these compounds.

Catechins

Tea, especially green tea, is particularly rich in catechins, a type of flavonoid that is thought to be responsible for many of the health benefits ascribed to tea. It is arguably one of the best studied components of tea. Catechins have proven antioxidant properties, have been linked with cancer prevention, lowering cholesterol levels, weight loss and may play a role in preventing diseases like Parkinson's and Alzheimer's. Despite these many potential benefits found in lab studies, there are limits as to how useful catechins may be in practice. Catechins easily bond with other components (such as caffeine) that then reduce their bioavailability, i.e. the amount that enters your circulation and is able to have an effect. They can also interact with proteins, and with iron, and inhibit the absorption of food proteins and iron in the body. Catechins are also unstable and easily degrade in poor storage conditions and at high temperatures (above 80°C), which is why you have to be careful when brewing green tea if you're to maximise its health benefits.

Methylxanthines (think caffeine!)

Tea contains a number of methlyxanthines, most notably caffeine, and theobromine and theophylline in much smaller quantities. Many studies have looked at the effect of caffeine and health. We've previously mentioned its downsides, and mentioned how it may bond with catechins and reduce their availability, but caffeine has also been linked with a host of beneficial health effects, from improving mood and cognitive performance, reducing anxiety, stimulating the central nervous system and cardiac muscle, and improving sports performance. Theobromine is thought to widen blood vessels and act as a diuretic. Theophylline is a known diuretic and central nervous system stimulant but also relaxes the lung smooth muscle, amongst other properties. It is a proven bronchodilator and is used alongside other medicines to treat asthma, bronchitis, emphysema and other lung conditions. Both caffeine and theophylline have been found in clinical trials to be useful for the treatment of premature babies, and for asthma, for pain relief and as a diuretic.

L-theanine

L-theanine is a pretty unique amino acid that is mostly found in the tea plant. It contributes to the 'umami' taste in tea. You might have heard this term in relation to other foods as it has been much talked about since being recognised in recent years. It refers to the fifth taste we sense with our tongue, along with sweet, salt, bitter and sour. It is a rich, savoury taste and is also found in foods such as cheese, tomatoes and meat. While L-theanine is linked to the aroma and taste of tea, it's also associated with health benefits. As mentioned previously, it has been shown to be associated with relaxation; it appears to facilitate the generation of alpha brain waves and the neurotransmitter, γ-aminobutyric acid, which are linked with a relaxed but alert mental state and improved ability to learn. L-theanine has also been linked with cancer prevention, the treatment of Alzheimer's disease, regulating blood pressure and weight loss, among other benefits.

Fluoride

Although it's not a major component of tea, fluoride is present in a big enough quantity to have an impact. For your average tea drinker, the presence of fluoride may have a beneficial effect on their teeth as it is known to help prevent tooth decay (there's more on this in Waters, Chapter I). However, a small number of people might be adversely affected if their fluoride intake is already high. The Republic of Ireland is a case in point. Not only do most people there receive a fluoridated water supply, they also consume a huge volume of black tea and this potential for excessive fluoride consumption has led some researchers to call for the implementation of risk reduction measures, namely to reduce the fluoridation of public water supplies. These researchers believe that the habitual tea drinking in Ireland may be contributing to a range of illnesses that could be lessened or avoided by reducing fluoride consumption. Another, admittedly extreme, example of the detrimental effect of fluoride in tea comes from the case of a forty-seven-year-old woman in the US who suffered chronic bone pain and had all her teeth removed due to their brittleness. She had developed skeletal fluorosis, a condition that is rare in places without high fluoride concentrations in the water. It turned out that for the previous seventeen years she had consumed a jug of tea made with 100–150 teabags every day! This gave her an estimated daily fluoride intake from the tea alone of more than six times the recommended amount.

Storm in a teacup? A health warning
You have to bear in mind that much of the research into these individual components of tea is undertaken in the lab, not in humans, and doesn't reflect how people actually consume them, how the compounds interact with other aspects of our diet and physiology, and to what extent they are really having a benefit. In addition, many of the effects will result from combinations of compounds working together. So, while on the surface it seems that we could avoid cancer, Alzheimer's and obesity, for example, from upping our tea consumption, in reality we know things aren't so simple. We all know people who have suffered these conditions despite their significant tea intake over decades.

Brewing temperature, brewing time, tea type and quality all affect the amount of beneficial compounds that are drawn out into the tea we drink, as well as how much is ultimately bioavailable to have an effect (i.e. how much taken in actually gets into our cells to do something useful rather than just being quickly broken down and peed out again).

> **It's a matter of timing**
>
> The time of day tea leaves are picked can affect their quality. The amount of starch tea leaves contain contributes to their quality. Starch production in tea occurs at dawn and dusk and leaves harvested in the morning typically contain more starch and are considered to be better quality than leaves picked later in the day.

Bearing in mind the caveats, tea appears to be full of beneficial compounds and regular consumption of these may well be doing us good in some way. But let's now take a quick look at what science tells us about the benefits of tea as a whole.

Tea and health
While tea extract and particular compounds have been found to be beneficial in animal studies, the health benefit in humans of tea as a whole seems inconclusive. That's not to dismiss this wonderful beverage, however, as there is promising research out there.

Some work into the benefits of tea has suggested that it may lower your risk of type 2 diabetes. However, many of the findings are from studies in mice and research in humans has led to inconsistent results. A meta-analysis[4] by a team from China looked into this in 2014 and found that while research findings have indeed been inconsistent, there seems to be an association between drinking three or more cups of tea per day and a lower risk of type 2 diabetes. Unfortunately, this analysis was unable to determine differences between the type of tea consumed by the research participants.

4 A type of study that combines the data from multiple studies. This approach is a more rigorous and reliable way of interpreting research findings than stand-alone studies.

Ultimately, studies looking at whether tea consumption may reduce the risk of a host of conditions, such as cancer, heart disease, dementia, diabetes and arthritis, to name a few, have produced inconsistent results in humans. It seems there are too many variables that make it challenging to pin down the health effects of tea: the type and quality of the tea, how it is brewed and prepared, how many cups per day a person consumes and for how many years, whether it is consumed with meals or without, and an individual's level of health – all these things may impact upon research results.

Research has delivered convincing evidence that tea, of various varieties, contains compounds likely to benefit health. How much of these make it into our cuppa, are absorbed into our bodies and in a way that can be used for benefit is still somewhat unknown, however. Overall, I think we can probably agree that tea is a good thing, but can you have too much of a good thing? Well, we know that excessive consumption could lead to too much fluoride, and too much caffeine might be a problem for some, but a few cups a day probably shouldn't cause too much of a problem. Is there anything else of concern? Something a little more sinister could, in fact, also be lurking in our tea. A number of tea studies have identified the presence of metals, such as aluminium, lead, manganese, cadmium and copper, which are picked up from the environment in which the tea plants grow and the water in which tea is brewed. But before you panic, the levels are generally low and, for the most part, it's not thought that moderate tea consumption, brewed for three minutes or so, would expose the general population to harmful levels of these elements. There may be some population groups, more vulnerable to exposure to metals, who might limit their tea intake as a precaution.

Black tea
Black teas are the most commonly consumed teas. There are many types and they are usually either from a single place of origin (e.g. Assam, Darjeeling, Ceylon, Lapsang Souchong, Turkish) or a blend

of teas (e.g. English Breakfast Tea, Afternoon Tea, Russian Caravan Tea). Some tea blends are also flavoured, such as Earl Grey (flavoured with the oil from the rind of bergamot fruit – I've tried this wonderful citrus fruit and it has a uniquely floral yet zesty taste; you should give it a go) and Lady Grey (Earl Grey tea additionally flavoured with orange and lemon peel). Ever asked for a cup of chai tea? Yes? In that case, did you know that what you're actually asking for is a cup of 'tea tea'? This is because chai is the Hindi word for tea. Chai, as we might order it in a coffee shop, is a beverage that is made from black tea blended with a mix of spices, sugar and milk. As you might imagine, the blends involved vary widely from place to place. It is reportedly quite different from traditional Indian chai.

While green tea tends to get the headlines in terms of its benefit to health, black tea has also long been studied for its effects. An analysis of a range of studies looking at the effects of black tea on health concluded that there is convincing and sufficient evidence to show a link between tea intake (of at least three cups per day) and a reduced risk of coronary heart disease. The study was published in a reputable publication (*European Journal of Clinical Nutrition*) but it should be flagged up that it was sponsored by The Tea Council and may not be considered entirely without bias. A separate meta-analysis by a different research team found that while black tea consumption didn't have an effect on total cholesterol levels, it was associated with lowering LDL[5] cholesterol – which is sometimes referred to as the 'bad' cholesterol. However, another meta-analysis in 2014 also found that studies into this area were inconsistent in their findings and concluded that, on the basis of their results, it is unlikely that black tea is having a significant effect on our cholesterol levels.

One project found no harm from drinking black tea but it recommended that people keep their consumption below eight cups a day to avoid potential adverse effects on iron status. Other research has

5 Low density lipoprotein.

found no significant effect of black tea on iron levels in most people. In those who already have low iron levels or are at risk of iron deficiency, it has been recommended that they limit their tea drinking between meals to reduce the possibility of it inhibiting the absorption of iron from their food.

An Australian team demonstrated that regular consumption of black tea (i.e. three cups per day over six months) lowered the blood pressure of people with normal to high blood pressure. Not all studies that have looked into this potential benefit of black tea have found such a benefit and it's again worth pointing out that this study was part funded by Unilever, a manufacturer of many of the world's most popular brands of black tea.

Green tea

Green tea is most popular in China, Japan and Korea but has been gaining popularity in recent years in many other parts of the world. Jasmine tea is tea, usually green tea, that has been scented using the blossoms of the jasmine plant. Jasmine blossoms are placed on top of dried green tea for several hours to release their aroma.

Green tea has long been known to be chock-full of advantageous compounds, particularly catechins, and has been linked to a host of health benefits. Some research in humans has found that those who regularly drink green tea may have a lower risk of a number of cancers (not all studies have found such an effect) and lower risk of dying from cardiovascular disease or heart attack. Work into the anti-inflammatory effects of green tea is yielding positive results and some research has even found that consuming green tea is linked with experiencing fewer bouts of flu and fewer cold and flu symptoms. The effect of this beverage on lowering blood pressure has also been shown. In addition, it's been found that green tea may help reduce anxiety and improve memory and attention, amongst other cognitive benefits. There is even evidence to suggest that green tea catechins are helping our dental health by inhibiting the formation of a bacterial biofilm in our mouths (i.e. they reduce the ability of bacteria in our mouths getting together in a film and sticking to our teeth). This action helps to keep our teeth and gums healthier. However, not all areas of research into the benefits of green tea have found an effect. Although some people believe green tea to be helpful for losing weight, a Cochrane Systematic Review of fourteen randomised controlled trials found that there was no significant difference in weight loss between those who took green tea preparations and those who didn't and, as such, green tea is unlikely to help you lose weight.

Many of these studies are, however, observational at this stage and can't prove that green tea is providing specific protection against disease and, while the evidence is mounting, the data so far are fairly contradictory and inconclusive. Further research is required to consolidate the findings of all these studies and offer more convincing evidence of the benefits of green tea. Interestingly, many white teas have similar levels of catechins and other polyphenols as green tea and may have similar health benefits, despite less research being undertaken on these teas.

Too much of a good thing?

In 2014, Jim McCants, from Texas, US, discovered that too much green tea might not be so good for our health after all. Wanting to get healthy in mid-life, he started to take green tea supplements for their supposed cardiac benefits. After around three months he became acutely ill and was given the news that he needed a liver transplant. The presumed cause of his liver damage was the supplements. Mr McCants isn't the only one to have experienced such ill-effects; green tea extracts have been implicated in tens of other cases of liver injury. It is not entirely clear what exactly in green tea extract is causing the

---

problem but it seems it may be related to a type of catechin called epigallocatechin-3-gallate (EGCG). EGCG has been found to lead to toxic effects in animal studies. According to the EFSA, while drinking green tea is no cause for concern, extracts used in supplements have the potential to result in ill health and, for some, the effects could be serious.

You may have come across something called matcha. It is quite the rage and hyped as a superfood but what exactly is it? Matcha is the dried, powdered leaves of green tea grown in Japan. The tea plants used for matcha are grown slightly differently from regular tea plants – they are covered for the few weeks before harvest, which makes the leaves very fine and very green as the amount of chlorophyll increases. This method supposedly helps control the tea's bitterness while giving it a full body. Matcha powder can then be made into a frothy tea drink or used as an ingredient in the preparation of a wide range of foods, from cakes to ice cream to granola, candies and smoothies. What makes matcha so amazing and is it particularly good for us? One of the reasons that matcha tea is thought to be superior to a cup of regular green tea is that you actually consume the leaves, whereas they are left behind in regular tea. In theory, this means that you must be taking in far more beneficial compounds. However, although you might be consuming them, it doesn't mean they are bioavailable in a similar way as in regular tea or that they confer any additional benefits. In fact, there's very little independent scientific evidence so far to show a significant health benefit of matcha tea over regular green tea, and certainly nothing convincing for other products made with matcha. With its distinctive and acquired taste and no proven additional health benefits, let's just say that I probably won't be introducing matcha powder to my baking just yet.

What are herbal infusions and tisanes?

There are many other drinks that we might refer to as tea that are not actually tea. Unless they derive from the *Camellia sinensis* plant they are technically tisanes – i.e. an infusion made from something other than tea (e.g. mint, chamomile, rooibos, fruit 'teas'). Tisanes may be made from the leaves, flowers, seeds, bark, roots and fruits of a range of plants.

Rooibos, or 'redbush', is a widely drunk tisane. It hails from the *Aspalathus linearis* shrub, indigenous to South Africa. In recent years, its popularity beyond southern Africa has grown enormously possibly in part due to the bestselling *The No. 1 Ladies Detective Agency* book series, by Alexander McCall Smith, in which the wonderful Precious Ramotswe frequently extols its virtues. Rooibos is naturally caffeine-free, like many herbal infusions, lower in tannins and less bitter than green and black tea. It is claimed that rooibos is a good source of antioxidants, a whole host of celebs have raved about its medicinal powers, and practically all packaging of rooibos boasts of its many healthful properties. Could this be the super-drink we're after? In a word, no. Unfortunately, there is little scientific evidence to demonstrate any specific advantages of this drink. While rooibos leaves contain plenty of polyphenols to start off with, these are significantly reduced by the processes that prepare them for brewing, i.e. traditional[6] fermentation, sun-drying, sieving and steaming. And with this reduction comes an associated reduction in any anti-cancer and antioxidant properties. It seems that many studies that have found a positive health effect of rooibos used an extract of the leaves (typically made by freeze-drying and powdering the leaves), at which point they contain higher concentrations of

6 Traditional fermented rooibos is the form that most of us are likely to buy but you can also get green rooibos, that is unfermented. Green rooibos appears to contain higher levels of antioxidants although any additional health benefits are still to be proven.

polyphenols, and not the brewed drink itself. Rooibos may prove to have particular health benefits when consumed as a herbal extract but as a drink the findings are currently unconvincing. We still need robust trials comparing any effects on those drinking cups of rooibos tea with those who don't drink it to demonstrate a genuine advantage of this beverage. This is all much to my chagrin as it's my 'tea' of choice. I start every day with a cup of redbush – I might as well just drink a cup of hot water.

While *bona fide* teas have been oft studied for their health properties, tisanes have had less attention and there is a considerable dearth of scientific evidence on any potential benefits. Research has unearthed a range of compounds contained within different herbal infusions that show potential, such as antimicrobial properties of peppermint tisane and anti-inflammatory properties of chamomile tisane, but rigorous studies into the effects in humans have not yet been sufficiently conducted. Furthermore, we need to be critical in separating out those studies that use plant extracts from those that investigate the drinks themselves, as this makes a big difference in the levels of compounds contained within them and the potential to have health benefits. It may turn out that to really benefit we should be chomping down supplements rather than boiling the kettle for numerous brews.

Taking Instagram by storm, with celebrity endorsements and the promise of a quick health fix, 'detox teas' have become the health drink *du jour*. (This is, of course, at the time of writing – trends like these move on pretty quickly and it'll probably be something else by the time you're reading this!) Detox teas are typically herbal blends that claim to cleanse the body from within, improve digestion and nourish the liver, boost the immune system and charge the metabolism, helping to promote weight loss, eliminate bloating, cleanse your skin and generally improve health. There is a particular focus on weight loss and a flat tummy for many of the products. There are myriad types, containing different combinations of ingredients, such as peppermint, chamomile, ginger, rooibos, rose petals, senna and actual tea (e.g. green tea, oolong tea). They can have diuretic and laxative effects and often contain quite a lot of caffeine, so may feel like they are physically doing something, but in the longer term these teas have not been found to have any beneficial health effects. The concept of detoxing may be popular but it is repeatedly called out by experts as nonsense (see *What are wellness drinks?* in Chapter 4, Cold drinks for more on detoxing). Worse still, some doctors warn that these teas may be causing harms, from digestive issues, dehydration, interfering with the effectiveness of contraceptive pills and promoting eating disorders. These problems are largely due to the inclusion of the laxative senna in the teas.

What is yerba mate?
Yerba mate is a drink, described as tasting a little like a smoky green tea, that is particularly popular in South America. In fact, it is considered the national drink of Argentina, Paraguay and Uruguay and apparently there it's more popular than coffee, tea and chocolate combined. It is highly regarded and often said to offer the 'strength of coffee, the health benefits of tea and the euphoria of chocolate' in one drink. Sounds amazing! Yerba mate is made from the leaves of the South American holly tree, *Ilex paraguariensis*. It can be prepared in a variety of ways, from a tea infuser or in a mug with a teabag, to a coffee machine or cafetière, or the traditional method with gourd and bombilla, and it can be served hot or cold, with a number of accompaniments (milk, honey, lemon, mint etc). It is naturally caffeinated and also contains theophylline and theobromine, stimulants found in tea and coffee. Unlike regular tea, yerba mate has a low tannin level so can be brewed strong without getting bitter. Yerba mate has been associated with a number of health benefits including counteracting obesity and inflammation. On the flip side, there has been some concern about a potential link between yerba mate consumption and cancer of the oesophagus, larynx and oral cavity.

Data suggest there may be an increased risk of these types of cancers in yerba mate drinkers but why this might be the case is unclear. Little is known as to whether there may be a link with the high temperature of the yerba mate when drunk or whether there is some constituent or constituents within it that may be causing a problem. What is clear is that a lot more work will be required to tease out what's actually going on. Whether beneficial or harmful, trials in humans with the beverage itself are required. Much like research into tisanes and some teas, the evidence is largely observational or from laboratory and animal studies, often using tea extracts.

A note on commercial iced tea and bubble tea
Many of us like to prepare a refreshing jug of iced tea on a hot day and have our own special recipes. Ready-made iced tea is also commonly found in the soft drinks aisle in some countries. On the surface it sounds revitalising and a healthier alternative to soda, but is it? Not really. As well as water and a tiny amount of tea extract (less than 0.5 per cent of the total), commercial iced tea typically contains sugar, sweeteners and flavourings, amongst other things. The bottom line is that commercial iced tea is unlikely to offer any health benefits over many other soft drinks. The same can probably be said for bubble tea. Bubble tea shops have popped up seemingly everywhere in the last few years. Hailing from Taiwan but hugely popular around the world, bubble tea is a drink comprising black tea blended with milk, sugar and fruit juice, with added chewy tapioca pearls. Served either hot or cold, it derives its name from both its tapioca balls and the frothy bubbles caused by vigorous shaking during its preparation.

What is coffee?
Prussian King Frederick the Great purportedly banned coffee drinking in 1777 as he believed that it was affecting the sales and consumption of beer. He stressed the superiority of beer over coffee,

proclaiming that he and his ancestors were brought up on beer, and many soldiers nourished on it. King Frederick referred to the increase in coffee drinking as disgusting. Now with coffee shops on just about every corner of the world, you can only imagine the horror he would feel if he were alive today!

Coffee is a hot beverage made by combining hot water and ground, roasted coffee beans. It is third in popularity in the world's favourite drinks and cherished for its stimulating effect. For many, a strong coffee is needed to get the day going; without it, they feel their brains are in a fug. This daily urge means coffee is big business. World production of coffee for 2017–18 was an estimated 159,663 million bags (at 60kg each). Finland tops the list of the world's biggest coffee consumers, at 12.5kg (dry weight) per person in 2016; in fact Nordic countries make up the top five, possibly reflecting their reliance on coffee to cope with the cold climate. According to the International Coffee Organisation, people in the UK pay the most for their coffee (at £11.45 per pound of soluble coffee in 2016, in comparison with £5.24 in Italy and just £2.23 in Poland, for example).

Coffee beans are derived from the *Coffea* plant. It's estimated there's around a hundred different species of coffee plant. They can be small shrubs to tall trees but the coffee plants typically used for the production of the coffee beverage are trees and they can grow up to 10m high. Coffee beans are the seeds of the tree's fruit, coffee cherries (technically berries). In the commercial coffee industry, the main coffee species used are Arabica and Robusta, with Coffee Arabica accounting for the greatest proportion. The three biggest producers of coffee are Brazil, Vietnam and Colombia. Coffee plants need particular environmental conditions to grow and the Arabica and Robusta plants have different requirements. While ideal temperatures range from 15 to 24°C for Arabica plants, the Robusta species flourishes in hotter conditions, between 24 and 30°C. Unlike Robusta that can be grown at relatively low altitudes, Arabica typically grows in hilly sites at higher altitudes and requires less rain than other species. Arabica coffee is generally sweeter than Robusta coffee and is considered to have a more complex flavour profile. It also has a much lower caffeine content (around half that of Robusta coffee). Robusta coffee is cheaper to produce as the plant is more hardy, easier to harvest due to its lower altitude and produces a greater yield. As a result, it tends to be used more for cheaper coffees, such as instant coffees.

Once the coffee cherries are harvested from the plants, the bean is separated from the outer hull. They are either passed through a pulping machine to remove the skin and pulp then washed and dried, or laid out in the sun to dry and then the beans removed from the cherries. The beans are then graded and sorted by size and weight. At this point they are called green beans and may be bagged up for selling. The all-important roasting comes next. It is during roasting that the aromas and flavours are developed. It's estimated that there are more than a thousand aroma components in coffee. You can vary the flavour profile of coffee by varying the roasting conditions. Coffee beans are roasted between 180 and 240 °C for anywhere between 15 to 20 minutes. As the beans brown, caffeol (an oil giving coffee much of its distinctive aroma and flavour) starts to emerge. The hotter and longer the beans are roasted, the stronger the aroma and flavour of the resulting coffee beverage. Lighter roasts tend to emphasise the different flavours of the coffee bean itself, with nuanced aromas and low bitterness. They have higher acidity levels and are more sweet than darker roasts. The darker roasts emphasise the roast character and are low in acidity, high in bitterness and less nuanced in flavour. Medium roasts fall somewhere in between with a defining characteristic of full body. The strong flavour of darker roasts might lead you to believe that they also give a greater kick from a higher caffeine content but the opposite is true. Lighter roasts actually have a slightly higher caffeine content.

Roasted coffee beans are ground to the size appropriate for their use (e.g. espresso machine, filter, cafetière, instant coffee), increasing the surface area and allowing the flavours and other compounds to be extracted more easily. Instant coffee is made by brewing and drying ground coffee. The coffee liquor is dried by either freeze-drying (where the liquor is frozen, cut into granules and then dried at a low temperature) or spray-drying (where the liquor is sprayed into a stream of hot air and as the droplets fall they transform into a fine powder that is then formed into granules).

The art of brewing coffee

Despite the heading, there is no particular art of brewing coffee because it all very much depends on how you like it. Having said this, there are a few things that you can do to get the most out of your coffee.

Clean equipment and fresh water are the first priorities. Leftover coffee grounds and grubby water filters in coffee machines and cafetières, for example, can result in a build-up of caffeol and limescale that affect the taste of the resulting coffee. Of course, your cup of coffee is only as good as the quality of coffee it is made from. Choose the coffee bean variety or blend, and roast type to suit your taste. Using coffee soon after it has been roasted results in a better taste. Similarly, using freshly ground coffee is best for an optimal cup of coffee. Buying your coffee beans or grounds in small quantities regularly may be the way to go to ensure it is as fresh as possible. Coffee needs hot water to ensure it is extracted properly during brewing but it doesn't like boiling water that can impair the flavour. Brewing time is also important. It needs to be long enough to dissolve the flavour compounds but not so long that too many extracted compounds lead to a bitter beverage. Brewing time may depend on the equipment used.

Coffee can be made via a range of apparatus, from a filter machine, stove-top espresso maker (or moka pot) or cafetière (or

French press), to aeropress or Pour Over. A filter machine does exactly what it says on the tin – water filters through the coffee to produce a drink. Coffee filters are often in paper form, made from a pulp of softwood that has a latticework structure, trapping the coffee grounds but allowing the liquid to pass through. The filter is placed in a coffee filter machine, coffee grounds are placed in the filter, water is added to another section, the water boils and flows through the coffee picking up the oils, aromas and other compounds on its way to produce a coffee drink. Stove-top espresso makers are divided into three parts: the bottom part holds the water, the middle chamber holds the coffee grounds and the top part receives the final product. As the water is heated on the stove and begins to boil, the steam is forced upwards through the coffee grounds. Extracted coffee flows up through a spout, spraying onto the ceiling of the top chamber, and collects in this chamber. In a cafetière, hot water is poured onto ground coffee and the coffee infuses, much like tea in a teapot. When the desired time/strength is reached, a plunger is pressed down, trapping the grounds at the bottom and allowing the coffee to be poured out without the grounds. The aeropress is a relative newcomer in the field of coffee making. Ground coffee sits in a chamber, hot water is added to it and it is stirred for a few seconds. A plunger then pushes trapped air down the chamber, forcing water over the coffee grounds and through a filter into your cup. It is considered an easier way to create an espresso although it is also used to make a longer cup of coffee. The Pour Over is also fairly self-explanatory. You place coffee grinds in a filter and pour over the hot water. Coffee will slowly drip through into a container. This is much like the filter machine but with subtle differences, including being able to control the water temperature with a Pour Over rather than relying on the automated settings of filter machines.

With all this choice, when it comes to making coffee can the workman blame his tools? Coffee aficionados all have their preferred methods and tools for making coffee, very much depending on their particular tastes. A better way to look at it is that there are many ways to make coffee and you should be able to find something that suits you.

Preparing that cup of joe

Before we go any further, why is coffee sometimes called a cup of joe? Inconveniently for our purposes, the origin of this phrase is uncertain but there are a number of theories, mostly of US origin. One is that the 'joe' may refer to the ordinary person, along the lines of 'your average joe', and that a cup of coffee is what the ordinary person would typically drink, hence 'cup of joe'. Linguists suggest that 'joe' is a shortening of Jamoke, which is itself a blend of java and mocha, a common nickname for coffee in the 1930s. It's possible that 'cup of Jamoke' simply became shortened to 'cup of joe'. Another idea postulated is that the phrase refers to Josephus Daniels, US Secretary of the Navy, who banned US ships from serving alcoholic drinks in 1914. The ban led to sailors consuming more coffee and the increased consumption was attributed to Josephus and the phrase subsequently coined. However, many refute this suggestion and highlight many holes in its argument. There are other theories but whatever the origin, the cup of joe phrase has been around for many decades and has stuck.

We've covered a range of ways to make coffee at home but across the globe, a great profit is made from us buying our coffee on the go. These days, coffee is prepared in a huge variety of ways, each with their own quirks and fan base. For some, it can be overwhelming to order a simple cup of coffee when presented with numerous coffee options and accompanying embellishments (skinny soy cinnamon dolce latte, anyone?). Let's take a brief look at the differences between some of the main coffee styles found in the biggest coffee shop chains.

Many of the coffees we buy are based on espresso shots in combination with other ingredients. As a reminder, an espresso is a strong black coffee made by forcing steam through dark roast coffee grounds. A sort of concentrated coffee shot. An Americano is made by simply adding extra hot water to an espresso shot. It was allegedly named after American soldiers in Italy during the Second World War who diluted the local espresso to make it taste more like what they were used to. One of the most popular coffee shop coffees is the Cappuccino, a drink of three layers: a shot of espresso, steamed milk and topped with milk foam. It may be then finished with chocolate powder or shavings. In Italy, it is drunk at breakfast time, or at the very least in the morning and tourists are often advised not to try and order a cappuccino after 11am, when it is much frowned upon. Caffe Latte (the most frequently purchased coffee type in the UK in 2017–18) involves steamed, frothy milk mixed with a shot of espresso, whereas a Flat White, particularly popular in Australia and New Zealand, pours 'flat' creamy, rather than frothy, steamed milk over an espresso. A Machiatto is similar to a cappuccino but it is a smaller drink lacking the steamed milk; foamed milk is added directly to the espresso. Caffe Mocha is a variation of the Latte and involves the addition of chocolate syrup, steamed milk and whipped cream to an espresso shot.

Coffee pods and capsules are another matter altogether. They're a bit like the teabag equivalent for coffee making. They both contain single servings of coffee grounds in a plastic container, covered with a foil seal. Coffee pods also bag the coffee grounds in filter paper. When placed inside a coffee machine, the foil breaks instantly and water that has been heated by the machine is forced at high pressure through the coffee and into your cup waiting patiently below. This is similar to a traditional espresso machine. The ease of coffee making and lack of fuss these machines offer has led to a huge rise in popularity. They provide consistent results every time and offer some of the coffee types found in the coffee shops but at a fraction of the price. However, the individual wrappings are wasteful and not typically environmentally friendly, the pods/capsules are more expensive than other methods of making coffee at home and you can't control the grind size.

Around the world, there are many alternative preferences for how coffee is made. To prepare Turkish coffee, finely ground coffee is simmered, sometimes along with sugar and/or spices, in a small brass or copper pot, a cezve. The result is a strong black coffee with a rich foam on top. As there is no filter, the coffee grounds are also contained in the cup but they fall to the bottom and are, for the most part, not consumed. Yuanyang (also known as Kopi Cham) is a Malaysian coffee drink, served hot or cold, that comprises black coffee blended with milky black tea. Vietnamese Ca Phe Trung is a coffee made with egg yolks, sugar and condensed milk. In Mexico, the Café de Olla is a traditional spiced coffee made with cane sugar and a cinnamon stick in a clay pot. Spices are also used in popular coffees found in places such as Morocco, Senegal and Saudi Arabia. Alcohol is another popular addition to coffee. The Irish coffee involves a splash of Irish whiskey along with cream and sugar, whereas the Pharisäer coffee in Germany replaces the whiskey with rum and puts chocolate shavings on top. In Portugal, a popular iced coffee is the Mazagran, consisting of espresso and lemon juice or lemon soda.

What is in coffee?

As well as all those aroma compounds and caffeol, what else is in coffee? The stimulant caffeine and other methylxanthines are the first, and best known, coffee compounds that spring to mind (covered in *What is caffeine?*). Beyond methylxanthines, chlorogenic acids (CGAs) account for over 10 per cent of the dry weight of green, unroasted coffee beans although many degrade during the roasting process. These antioxidants are responsible for much of coffee's bitter taste and can cause acid reflux in some coffee-drinking sufferers. As CGAs break down in the gut, they may help stimulate gut bacteria, in effect acting as prebiotics, and they may also have anti-inflammatory properties. Research into this area is at an early stage but, with the current trend favouring gut environment (or microbiome) research, any effects from coffee on this part of the body will be eagerly investigated.

Diterpenes, mainly cafestol and kahweol, are oily substances in your coffee. The evidence for their effects is currently mixed. Some research has shown that these compounds increase cholesterol levels, particularly the so-called 'bad' cholesterol LDL (low density lipoprotein) but the good news is that the effect was limited to those who drank unfiltered coffee. Unfiltered coffee includes coffee made in a cafetière or Turkish coffee, for instance, but most people consume filtered coffee. It seems these diterpenes are trapped by the filter. However, diterpenes may have beneficial effects (e.g. anti-cancer properties, although such research is too early to confirm this benefit) that are being lost through the filtering process.

Trigonelline, which comprises around 1 per cent of coffee by weight, plays a role in the development of flavour compounds during roasting. Many trigonelline compounds are destroyed by the heat of the roasting process and lead to the formation of pyridines, amongst other things. Pyridines help to produce the sweet and earthy aromas often found in coffee. Another by-product of the decomposition of trigonelline is niacin (also known as vitamin B3). Niacin is used by the body for turning food into energy and is important for keeping your nervous system, digestive system and skin healthy. A cup of Americano-style coffee can contain somewhere between 1 and 3mg of niacin. As recommended daily allowances of niacin are 14mg for women and 16mg for men, if you have two or three cups of coffee a day you can see that this beverage is providing a useful source of this essential vitamin. Trigonelline itself may have an additional health benefit. Some research has found that it may help prevent dental caries as it appears to inhibit the bacteria *Streptococcus mutans* from sticking to teeth, a critical feature in the development of tooth decay.

Talking of teeth, it's not just tannins in tea that result in staining. Coffee drinkers also experience staining of their teeth. While black tea is a worse culprit with its high tannin content, coffee still contains compounds that leave their mark on our smiles. Coffee is full of chromogens, containing dark pigments (imparting the dark colour you see in coffee) that cling to tooth enamel. Tannins can bind to the chromogens and make them more sticky. While you may reduce the staining of tea by the addition of milk, which binds to the tannins, chromogens still stain your teeth even with the addition of milk in your coffee. But adding milk will bind up some of the tannins, lessening their direct staining effect, as well as reducing their availability to bind to the chromogens and enhancing their stickiness.

Coffee: the cockroach's favourite drink?

Along with hot water, there are many, many chemicals that go into making your cup of coffee. One of these, 2-ethylphenol, found in Arabica coffee, contributes to coffee's aroma, but it also acts as a pheromone for cockroaches, helping them chemically communicate. The jury is still out on whether cockroaches are particularly attracted to coffee – some say yes, others no – but in warm climates they are often found living in coffee machines. Consider yourself warned!

With the wide array of coffee preparation, brewing methods and serving styles, it will come as no surprise that concentration levels of the various compounds in coffee vary widely in the end cup.

Coffee and health

Regardless of the effects of individual constituents of coffee, there's no doubt that we're constantly being told that this drink is good for us (albeit minus the creams and syrups). So is there a solid basis for this conclusion?

First of all, a lot of observational studies have found that regularly drinking coffee reduces your risk of dying prematurely. For coffee drinkers, this must be music to their ears! Observational studies have also led to the conclusion that it is probable that coffee decreases the risk of developing some types of cancer. Some people have been concerned that regular coffee consumption may have a particular adverse effect on the heart but science does not support this. There's no hard evidence to demonstrate that coffee increases the risk of coronary heart disease, congestive heart failure or sudden cardiac death, with moderate intake. There's pretty good evidence from meta-analyses to suggest that coffee may even reduce the risk of cardiovascular disease, Parkinson's disease and stroke, and prevent the onset of type 2 diabetes, liver disease and gallstone disease. The evidence confirming the benefit of coffee in lowering the risk of developing type 2 diabetes is particularly strong. Studies involving well over a million people have shown that regular coffee consumption may reduce the risk by as much as 30 per cent.

What about the brain? There's quite a lot of research into the effects of caffeine on the brain but here we're interested in the effect of coffee as a drink. There have been reports that coffee can help us improve our mental performance in the short term and even prevent cognitive decline or dementia in the longer term. Starting with the short-term effects, the evidence is conflicted with very inconsistent results across studies, in different populations and with different measures of cognitive function. Some observed better performance in cognitive/memory tests in regular coffee drinkers (e.g. in older women), whereas others, despite their efforts, found no significant association between coffee and cognitive performance (e.g. in older men).

Wake up and smell the coffee

An unusual study found that people performed better at an aptitude test, often required by business schools, after being exposed to coffee aromas (not even drinking it), versus those not exposed to any scent. The research tested around a hundred business students, and followed up with a questionnaire about their performance, and found that participants in a coffee-scented environment did better at their analytical task and had heightened performance expectations. In other words, they believed they would do better in their analytical reasoning as they associate coffee with increased physiological performance, such as alertness. The researchers conclude that coffee aroma can produce, in effect, a placebo effect on performance. It makes you wonder whether it's worth placing a strong cup of coffee beside you when you're trying to undertake a focused task, even if you don't drink it.

Regarding longer-term effects, looking at research to date, there are signs that drinking coffee regularly is associated with a lower risk of cognitive decline, dementia and Alzheimer's Disease. However, the data is not robust and the evidence is still inconclusive. We need to be cautious in our interpretation of the effects of coffee on the brain until further, well-designed, large trials have been conducted.

Caffeine has long been linked with increases in physical performance. Research in this area often focuses on athletes, but there is evidence to show that caffeine can also improve performance in those who are not trained athletes. That's probably not that surprising but can you get the same effect from coffee as a drink, not just from the stimulant compound caffeine? There's moderate evidence to show that coffee can help endurance cyclists and runners better their performance. A number of studies found performance improvements in those who drank coffee at least 45 minutes before undertaking endurance tests.

So, coffee appears to have a range of health benefits but is this really just all down to caffeine? Would an energy drink do just as well? After all, many of the stories we read in the news about health claims are all about caffeine. Perhaps that's really the wonder-drug here. While we mainly associate coffee with caffeine, it seems that its health benefits are actually due to a wider collection of compounds. Many of the health effects seen with coffee consumption have also been observed in those who drink the decaffeinated version, indicating that it is probably not caffeine alone that is responsible for coffee's beneficial effects but myriad other compounds. This is a good reminder that reading about the effects of coffee is not the same as reading about the effects of caffeine – they are not interchangeable (tabloid health journos take note!).

But not all research into the potential health benefits of coffee has revealed sound findings. If only we could drink a cup of coffee before meals to reduce our appetite and help in watching our weight. The possibility that coffee could help in controlling appetite, something that is commonly asserted, would be widely favourably received if it were true. However, studies to date have not resulted in convincing findings to support this. A whole host of other health conditions and possible links to coffee intake have been investigated but, as yet, there's a dearth of hard evidence to confirm any *real* effects one way or the other.

For some people there are also disadvantages of drinking coffee. High consumption of coffee has frequently been found associated with low birth weight, premature birth and pregnancy loss, which is why limiting caffeine intake during pregnancy is strongly advised by public health agencies. Also, bone health may be detrimentally affected by high consumption. Some studies found an increased risk of fracture in women who drank more coffee. The link between coffee and bone health is unclear as the same studies also found a lower risk of fracture in men who drank more coffee. It goes to show that individual responses to coffee vary widely and generalising population studies to individuals isn't very accurate.

But if we are to believe that coffee has beneficial effects, we also need to understand why. Why should coffee help our heart, lower our diabetes risk or prevent cancer? Understanding the mechanisms behind the relationship between coffee and health is very complicated and this is where much research is now being focused. When it comes to the health benefits, it seems that different components of coffee (e.g. caffeine, CGAs and diterpenes) may be exerting different effects, both on their own and in combination. The stimulating effects of methylxanthines may be key in improving sports performance, where it's thought that the caffeine in coffee may affect the production of adrenaline as well as change how we experience fatigue associated with physical exertion. The antioxidant and anti-cancer properties are likely exerting beneficial effects elsewhere, such as the liver. We still need to find out a lot more about what exactly is happening in our bodies when we drink coffee that leads to health benefits.

To sum up, the science seems to show modest evidence of health benefits of coffee and limited evidence of health harms for most people. But what you're probably interested in knowing is how much coffee you should drink to gain the most benefit while also avoiding potential harms. Well, for the average person who is not pregnant or at high risk of particular conditions, of the many studies that have identified benefits of coffee, it seems that the optimal amount is about three or four cups of coffee a day, versus none at all. Make of that what you will.

What is cocoa?

Chocolate, mmm. Just the thought of it has me salivating. (I know I'm not alone.) There's something so very perfect about its composition that makes us adore and crave it. Personally, I prefer the solid form but many of you are just as happy drinking it and this human need extends back numerous generations. Cocoa has been consumed by humans for centuries and was an important crop in ancient South American cultures. The Mayans, for example, created a cocoa-based drink for use in marriage rituals, and by the mid-1600s cocoa drinks were popular across France following their introduction to Europe by Spanish explorers. A Frenchman opened the first hot chocolate shop in London and such 'chocolate houses' became common in England by the 1700s.

Cocoa derives from the beans of the *Theobroma cacao* tree. This tropical tree requires hot, humid conditions and most of the world's cocoa beans come from West African countries. Despite the demand for this popular crop, the vast majority of cocoa is produced on small farms. Cocoa beans are actually the seeds of leathery cocoa pods, each of which contains around twenty to fifty 'beans'. When harvested, the beans are fermented in covered piles (the pulp layer surrounding the beans heats up and ferments the beans), dried, cleaned and packed up for processing into something more edible.

The beans are first roasted to bring out the chocolate flavour and colour. The temperature and time involved depend on the type of beans being roasted and the type of product they are being roasted for. The shell is separated from the nib (the contents of the bean) and the nib is ground into a paste. The grinding process creates heat, causing the cocoa butter in the nib to melt and creating cocoa liquor. Despite the name, this liquor is solid at room temperature. The liquor may be treated with an alkalising agent, such as potassium carbonate, to reduce its acidity. This makes it darker, milder and more chocolatey. The next step is to press the liquor to divide it into cocoa butter and solid cocoa presscakes. The butter is used in the manufacture of chocolate and the cakes are ground to form cocoa powder.

> **It amounts to a hill of beans**
>
> It takes around four hundred cocoa beans to make 1lb of chocolate. That's about one chocolate bar from each cocoa pod. Given that a cocoa tree produces about thirty to fifty pods a year and it takes between three and five years for each tree to bear fruit, you need a lot of cocoa trees to produce the vast quantity of chocolate that humans consume each year! We get through more than 4.5 million tons of cocoa beans annually.

What's the difference between hot cocoa and hot chocolate?

Hot cocoa is made from the cocoa powder whereas, strictly speaking, hot chocolate is made using dark chocolate where the cocoa butter is intact. True hot chocolate is therefore a richer beverage. To make these drinks, the powder or chocolate is combined with a range of other ingredients such as warm milk, hot water, sugar and flavourings. You can buy cocoa powder or drinking chocolate and add your own ingredients to prepare your drink but more instant versions are hugely popular.

Commercial hot chocolate drinks, like those you might buy in a café, or sachets of 'hot chocolate' powder you prepare at home, typically contain a host of additional ingredients. For example, here is the list of the ingredients in a Starbucks' 'Signature Hot Chocolate': milk, water, signature hot chocolate (sugar, cocoa powder processed with alkali, milk, cocoa butter, vanillin), whipped cream (cream, sugar, propellants: nitrous oxide, nitrogen; emulsifier: E471, flavouring, stabiliser: carrageenan)[7] . Make that a caramel version and you can expect the following further ingredients: caramel flavour syrup

7 Starbucks, 'Summer 2 2018 beverage ingredients.' https://globalassets.starbucks.com/assets/68FC43D2BE3244C9A70EE30EA57B4880.pdf

(sugar, water, natural flavouring, colouring food: concentrate of apple, carrot, hibiscus, molasses; acidity regulator: citric acid, preservative: potassium sorbate), caramel drizzle (sugar, dextrose, glucose syrup, butter, fructose, double cream, skimmed milk powder, natural flavouring, emulsifiers: soya lecithins, mono-and diglycerides of fatty acids; acidity regulator: sodium citrates, salt, stabiliser: triphosphates, antifoaming agent: dimethyl polysiloxane)[8]. That list is somewhat overwhelming to take in but why are all those ingredients included? As well as enhancing flavour, these extra ingredients offer a number of roles from thickening up the drink, improving its colour, keeping the dry product stable during storage and extending its shelf life, and helping to ensure the resulting drink is the same every time.

Do cocoa-based drinks have health benefits?
Any health effects stemming from cocoa-based drinks will very much depend on what else they contain and how they're prepared. (A large mug of sugary, synthetic hot chocolate topped with marshmallows and whipped cream is unlikely to deliver the same effects as a simple hot cocoa with no toppings.) To find meaningful benefits we need to tease out the effects of cocoa from chocolate products and other drink ingredients.

Cocoa powder contains a surprising amount of fibre (26–40 per cent of its composition), as well as proteins, carbohydrates, fats, minerals and vitamins. Sounds like the whole package! Cocoa also

---

8 As a comparison, Cadbury's 'Hot Chocolate Instant' powder, widely available in supermarkets, contains: sugar, whey powder (from milk), fat reduced cocoa powder, glucose syrup, vegetable fats (coconut, palm), skimmed milk powder, milk chocolate (milk, sugar, cocoa mass, cocoa butter, vegetable fats (palm, shea), emulsifier (E442), flavourings), thickener (E466), salt, milk protein, anti-caking agent (E551), flavourings, acidity regulator (sodium carbonate), emulsifier (E471), stabiliser (E339). (Cadbury, 'Hot chocolate instant.' https://www.cadbury.co.uk/products/cadbury-hot-chocolate-instant-11688)

contains the methylxanthines caffeine and theobromine. Despite their potential to benefit health, it is the richness of flavonoids in cocoa that has attracted most research interest. Observational studies have suggested a positive relationship between the consumption of cocoa and blood pressure. Flavonoids in cocoa have been linked to heart health benefits, assisting the arteries and helping maintain normal blood flow. However, most of these findings stem from animal studies and have not been satisfactorily demonstrated in humans. Antioxidant and anti-inflammatory effects of cocoa have also been indicated, sparking interest in its potential to help prevent cancer, but knowledge in this area is limited and robust clinical intervention trials are still needed to prove any such findings.

While cocoa may contain beneficial compounds, most of us are unlikely to consume them in sufficient quantities to have any real effect. I'm sure people would love to increase their intake if it were in chocolate form, but this would come with the addition of unhealthy fats and sugars, negating the benefits and increasing the potential harms. Cocoa is bitter by itself and increasing consumption of it alone is a bit of an ask for most of us. The populations who saw benefits to their blood pressure were small communities in different parts of the world drinking large quantities of cocoa, that is particularly rich in polyphenols (e.g. the Kuna Islanders of Panama, who have very low rates of hypertension, and who drink five cups of home-grown, flavonoid-rich cocoa per day). In addition, the cocoa that many of us consume has undergone many stages of commercial processing and is likely to contain far fewer beneficial compounds as some are lost at each stage. For example, fresh and fermented cocoa beans contain 10 per cent flavonols, a class of flavonoids, prior to processing, while the resulting cocoa powder contains about 3.6 per cent.

There hasn't been as much health research into cocoa-based drinks as tea and coffee so we don't yet have a body of evidence behind its effects. Results from research into the health benefits of cocoa have to date been varied but signs from a mix of observational studies, mechanistic work and interventions look promising. But even if there turns out to be no significant effect, we should probably focus on just how delicious such drinks are – sometimes there's more to life than the nutritional value of what we consume.

What are malted milk drinks?

Now we come to the comfort blanket of the drinks world: the range of drinks that have played a central role in the bedtime routines of thousands around the world, namely malted milk drinks. But what exactly is a malted milk drink? These are typically hot, milky drinks made with malted barley,[9] sugar and other ingredients. The word 'malt' comes from 'maltose', a type of sugar produced by the breakdown of starch and produced in milled malted grains when combined with hot water. In essence, malt is formed when barley is dried, submerged in water to stimulate germination, then dried again and heated in an oven. The germination is halted when sprouting begins, and when the starch content is at its peak, which makes the barley more open to the conversion of its starches into sugars. Malt can be further processed in a number of ways and is used for a variety of purposes, including in brewing (more on this in Chapter 5, Alcoholic drinks) and food products. Malt extract, often used in food products and malted drinks, is made by combining malt with hot water, digesting the starch into a mix of complex sugars and amino acids. This solution is then evaporated to create a thick, viscous extract. It may be used in this form or further dried into a powder. The end result is a malt that can impart a roasted flavour and colour, and sweetness, to malted milk products.

9 Other cereal grains may be similarly malted, such as oats, wheat and rice, but barley is the most commonly malted grain.

Three well-known brands of malted drinks are *Horlicks* (created in 1873 by the British Horlicks brothers, in their US company), *Ovaltine* (developed in 1904 in Switzerland by chemist George Wander as Ovomaltine, but upon exportation to Britain in 1909 a spelling error on the trademark application led to its current name) and *Milo* (devised in 1934 by Australian food scientist Thomas Payne, who drank it every day until he died aged ninety-three). You might be forgiven for thinking they are all the same thing but they are actually subtly different.[10] Horlicks is a powder of wheat flour, malted barley, sugar, milk, palm oil, salt and added vitamins that you mix with hot milk to make the drink. Ovaltine is a powder of malted barley, milk, cocoa, sugar, rapeseed oil and added vitamins and minerals. Milo is very similar to Ovaltine in its composition but includes tapioca starch and non-specified 'flavouring', replaces rapeseed oil with palm oil and contains a slightly different combination of vitamins and minerals. However, unlike Horlicks and Ovaltine, the Milo malted drink is prepared with hot or cold water, rather than milk.

Comforting, warming, soothing; all adages associated with malted drinks. They are identified with bedtime and assisting sleep but does the science back this up? Before considering whether they might help sleep, could these drinks in fact hinder sleep? After all, we know that cocoa, found in some malted milk drinks, contains the stimulant caffeine. Despite the addition of cocoa in some malted drinks, the amount is so small that any caffeine content is considered negligible, if non-existent. So malted milk drinks may be a good alternative to other hot drinks in the evening. However, there is a considerable amount of sugar in them (over five teaspoons per serving in Horlicks and Ovaltine, for example – imagine putting that much sugar in your tea or coffee). Sugar is quickly utilised by the body for

10 The ingredients listed are those for the UK versions of the drinks.

energy and can lead to a fast rise in blood sugar and insulin production. This effect may not be that helpful just before going to bed.

Moving on, before you pop on your slippers and reach for that jar of Horlicks for a good night's sleep, unfortunately there just isn't substantial evidence to conclusively demonstrate that malted milk drinks improve sleep. While various small studies suggest there may be sleep-inducing effects of these drinks, when looking at their individual components it seems that any such effects may be down to the warm milk or the added vitamins, i.e. nothing to do with the malt. The presence of the amino acid tryptophan is often cited as the reason why warm milk can lead you to a slumberous state. Tryptophan serves a number of purposes. As well as being essential for infant growth, for example, it is a precursor of both serotonin and the hormone melatonin. Melatonin regulates sleep and wakefulness, and serotonin is a neurotransmitter with broad physiological actions including modulating sleep and mood. However, milk simply doesn't contain enough tryptophan to have a significant effect. You would need to drink a vast quantity of it to really send you off to sleep. Milk does contain other elements that may contribute to the feeling of tranquillity; casein hydrolysate, that appears to have stress-reducing properties, and magnesium that may help reduce anxiety. Such properties may work together to help calm and soothe. As well as milk, certain vitamins, in particular B vitamins, have been associated with improved sleep, so their inclusion in malted milk drinks may be supporting their soporific effects. I say 'may', as the science behind all these possibilities is not strong.

There's a definite psychology behind malted milk beverages, persuading the consumer that these drinks will make them relax and feel at ease. This is half the battle won; you expect the drink to help you relax and so you're already in the right frame of mind to do so. Then there's the warm milk and richness of flavour and the association with childhood, being soothed with a warm milk before bed. And now, relax . . . zzzz.

Hot temperature: can it help or harm?

There's an old maxim that in hot weather we should drink hot drinks, rather than cold, because they encourage our bodies to release heat, thereby cooling us down. A saying is one thing but does this really work? Actually, there is some evidence to suggest. A small study by researcher Ollie Jay in 2012 found that while hot drinks added heat to the body, the body then reacted by significantly increasing its rate of perspiration to more than compensate and, as a result, cool the body. He found that so long as the sweat is able to evaporate, then hot drinks work well to cool you down. If the sweat can't easily evaporate (which might be the case in very sticky, humid conditions or if you're wearing a lot of clothing) then hot drinks will just make you hotter, so that's when to reach for a cold one instead.

On the flip side, is there any harm from drinking hot drinks? A good number of studies have investigated whether consuming hot drinks, and indeed other hot foods, may cause heat damage to the oesophagus and elsewhere in the body. One review analysed the findings of fifty-nine studies and found that higher drinking temperature of coffee, tea and mate was associated with an increased risk of oesophageal cancer. Another review found similar results, particularly regarding oesophageal squamous cell carcinoma. The link is thought to be due to the hot drinks causing thermal damage to the lining of the oesophagus. There are a number of methodological limitations to these studies but, nevertheless, the evidence available was sufficient for The International Agency for Research on Cancer, in 2016, to list very hot drinks as a probable carcinogen. Note that this refers to drinks of a temperature much higher than many of us drink – about 149°F.

Self-heating cans

Soon you'll be able to purchase (again) your tea, coffee, or chai latte in a can that heats up its contents for your convenience. The technology behind self-heating cans has been around for a long time, since the early twentieth century. The cans contain three compartments: one houses the drink, such as coffee; one holds water; and one holds a heating agent (the type varies between products). When the can is activated, for example by pushing a button on the bottom of the can or pulling a ring, the barrier between the water and heating agent (but not the drink) is broken and, as they come into contact with each other, this starts what's known as an exothermic reaction (a chemical reaction that releases heat or light). The drink absorbs the heat and warms up. This all takes around two to three minutes. In other versions, the can contains two different chemicals, rather than one plus water, which react when in contact with each other.

Originally, self-heating cans were designed to heat food for explorers and mountaineers (who have an obvious need for convenient, hot food), but the market expanded and soups and other foodstuffs in self-heating cans could be purchased in some parts of the world. Their popularity tailed off in the 1940s, although their use in camping consumables continued, but there has been a renewed commercial interest in them in recent years. In the late 1990s, food giant Nestlé tried to introduce self-heating cans of coffee but this venture failed to capture the market as customers complained that the coffee simply didn't get hot enough. It seems that these cans only add about another 40 degrees to the temperature of their contents, which is fine if the starting temperature is around 25 degrees but not great if it's much lower. And if you think about it, you're most likely to want a hot drink on a cold day so the base temperature of the can will probably be fairly cold. Getting the amount of reagent and the packaging just right is complex. In the early years, the cans could get too hot to handle and sometimes even explode. Other companies have also been busy investing in this technology, with enterprises in the US and Spain launching self-heating drinks, such as tea, coffee and hot chocolate, in 2018. Interestingly, the Spanish company, The 42 Degrees Company, initially utilised the technology to develop food and drink products for the emergency services, volunteers and victims caught in natural disasters or humanitarian crises, but subsequently decided that everyone should have the opportunity to enjoy their products.

CHAPTER FOUR

Cold Drinks (Non-Alcoholic)

Air accumulates in the stomach and increases gastric volume. This activates receptors in the gastric wall initiating a reflex that leads to the relaxing of the lower oesophageal sphincter, air moving upwards through the oesophagus and then passing through the upper oesophageal sphincter. It is then that it may be heard, and sometimes very loudly. What am I talking about? Burping, or belching, of course. Burping is essentially the physiological venting of excessive air from the stomach. Although it is a natural bodily process, many people around the world find burping socially unacceptable but there are a few places in which it is a sign of appreciation, such as Bahrain and parts of India and China. Many foodstuffs, illnesses and medicines can make us burp but fizzy drinks are an obvious culprit. Most people are likely to need to burp at some point, no matter how overtly or discreetly, after consuming a carbonated beverage. And soft drinks like these are incredibly popular, so at any one time there must be thousands of people around the world burping. Imagine if we could somehow capture and quantify the collective sound?!

Shop refrigerators typically overflow with the enormous range of soft drinks produced by manufacturers. They provide an interesting alternative to water and often boast of persuasive health benefits, but could they be doing us more harm than good?

Cold drinks: in essence

Soft drinks are not a modern invention. We may have been led to believe that everyone in the distant past drank beer or wine because the water was too dangerous to drink, but the truth is that many of the soft drinks we consume today were created hundreds of years ago. Barley water, that we often now buy as a type of dilutable fruit cordial, was originally simply a concoction of pearl barley and water and references to it go back as far as the fourteenth century. Unsurprisingly, herbal brewing to develop drinks also goes way back. While we're familiar with beverages like dandelion and burdock, and root beer, their origins lie many centuries before us when people, such as Captain James Cook when exploring New Zealand, brewed slightly fermented roots to cure ills (in Cook's case, brewing spruce beer to avoid scurvy). Of course, nowadays such drinks bear little resemblance to their original versions.

Soft drinks are so-called versus 'hard' alcoholic beverages. There are various definitions of soft drinks but here I'm using the term to refer to a broad range of sweet non-alcoholic drinks, typically served cold, that includes sodas, juices, flavoured waters, dilutable cordials, sports and energy drinks, and wellness beverages. Current trends in soft drinks are for those with more natural flavours and ingredients as well as functional additives – those added extras that may help us live longer with one easy gulp (or so we would like to believe).

We like cold drinks more than hot drinks (apparently)

We start off our lives drinking warm milk but, as we grow, humans seem to develop a preference for cold drinks. Cold drinks are preferred to hot for quenching thirst and the cooling effect in the mouth is generally perceived as pleasant. There's no definitive evidence proving why we're so drawn to cold drinks, but some think that we associate the cold sensation with the satiation of thirst and as refreshing. (It's worth noting, however, that this particular research was in part carried out by a prominent food and drink manufacturer, so they would be keen to promote these ideas.)

What is in soft drinks?

Soft drinks vary in their flavour and appearance yet they have many common ingredients. Water is common to all and is covered in the first section of this book. It's worth just mentioning here that manufacturers tend to make their soft drinks with soft water to avoid any distinct tastes from the water coming through. But let's take a look at what else might be included. The main ingredients of soft drinks tend to be a combination of sugars, fruit juice and additives, such as sweeteners, acids, preservatives and colours. Sodas also contain carbon dioxide.

Sweet talk

When it comes to soft drinks, the main ingredient that springs to mind, other than water, is sugar. Sugar is used in various forms in soft drinks. Sugar can be obtained from a range of plants, particularly sugar cane and sugar beet in the form of sucrose. Sucrose comprises two molecules bound together: one glucose molecule and one fructose molecule. Together with galactose, glucose and fructose are simple sugars, known as monosaccharides, that make up all carbohydrates. They bond together in different combinations to form complex carbohydrates. Regardless of their composition, once in the body all carbohydrates get broken down again into the three simple sugars.[1] Sugar is often added to soft drinks to provide sweetness. Sucrose, when combined with acid typically also found in soft drinks, transforms into an equal mix of glucose and fructose. Glucose and fructose syrups may also be included to contribute sweetness. The various sugars have different sweetness levels (e.g.

[1] As the simple sugars are monosaccharides, you may have guessed that sucrose is a disaccharide (i.e. comprising two monosaccharides). Other disaccharides are lactose (which is glucose plus galactose) and maltose (glucose plus glucose). Another term you may have heard is polysaccharide, which refers to more than ten monosaccharides linked together. Starch is an example.

fructose is sweeter than sucrose, which is sweeter than glucose) but they all provide the same amount of calories (4 kilocalories per gram).

As much as some health gurus might not like to hear it, we actually do need some sugar in our diet. Once your body turns your food into glucose, it is absorbed by your intestines and passes into the bloodstream. This is what is being referred to when people talk about blood sugar levels. Maintaining consistent blood sugar levels is important for our health. Your body produces insulin to help transfer the glucose to your cells for energy, whether for use straight away or stored for later need. Most of our cells require glucose to function, but particularly the brain. When it doesn't get enough glucose, the brain doesn't have enough energy to ensure efficient communication between the neurons and the rest of the body. This is why, in the short term, if you miss a meal or don't eat enough you may find it trickier to concentrate or remember things, and feel tired and irritable. Over a longer time, however, a glucose deficit to the brain can lead to serious consequences, such as cognitive impairments. However, too much sugar is also bad for our health. Sugar-sweetened beverage consumption is frequently linked in research studies to an increased risk of type 2 diabetes, dental decay, weight gain and obesity. There is also some evidence that higher consumption of these drinks is associated with non-alcoholic fatty liver disease and heart problems. Despite the need for some sugar, there's no doubt that most of us consume too much of it in our diet, so we don't really need to acquire even more from our drinks.

One ingredient of particular interest in soft drinks that has courted the headlines over recent years is high fructose corn syrup (HFCS). This syrup is derived from corn starch that is broken down into glucose molecules, and enzymes are added to convert some of the glucose into fructose. The amount of fructose in HFCS varies but the proportion is commonly more than 50 per cent fructose,[2] with the remainder being glucose and water. Our bodies treat fructose differently from glucose. Fructose doesn't have a significant effect on blood sugar or insulin levels, like glucose does, but it can have a more immediate effect on the fats in the blood (triglycerides). Studies have found that when high amounts of fructose are consumed, this is associated with higher triglycerides and fatty disease, decreased sensitivity to insulin and higher levels of uric acid, that causes gout. Some people are concerned that HFCS is in some way much worse than sucrose for health, due to its excess fructose content, as the metabolic pathway for fructose lacks some of the cellular controls that are present in glucose metabolism (in other words, fat production caused by the fructose may be unrestrained). However, studies to date have not found explicit, specific negative health effects particular to HFCS in comparison with sucrose. (It's worth noting that some of these studies have been funded by the soft drinks industry, who have a vested interest in using HFCS in their goods.) The whole area remains controversial but it's probably easiest just to assume that too much sugar of any type is not good for your health.

In the US, more than two in every three people (70.2 per cent at the time of writing) are either overweight or obese, and sugar-sweetened drinks are the single largest source of added sugar, and the top source of energy intake, in the US diet. With statistics like these being similarly replicated in many other countries around the world, public health professionals have been calling for action to reduce the consumption of sugary drinks. On 6 April 2018 the UK's Soft Drinks Industry Levy – commonly known as the 'Sugar Tax' – came into effect. The aim of the levy is to encourage manufacturers to reformulate their soft drinks to reduce their sugar content. The amount manufacturers have to pay is dependent on the amount of

2 These proportions largely refer to the US where HFCS is more commonly used in soft drinks.

sugar in their products, and in readiness 50 per cent of manufacturers reformulated their drinks prior to the tax coming into effect. The reduction in sugar content across these products was equivalent to 45 million kilograms of sugar every year. This all forms part of the government's goal of tackling childhood obesity and is in line with research that shows that reducing intake of sugar-sweetened beverages will reduce the prevalence of obesity and related ill health. Interestingly, Coca-Cola has stated that it would not reduce the sugar content in its classic cola. Instead it would pay the tax and pass the full cost on to retailers. To sweeten the bad news, however, it is reducing the size of some of its bottles (i.e. in effect charging the same price for less drink, presumably in the hope that consumers don't particularly notice) and at the same time increasing the size of some bottles of its no-sugar versions, Diet Coke and Coca-Cola Zero Sugar. They hope this will encourage consumers to purchase these varieties instead. The UK is no exception; Mexico, France, Ireland and Hungary also have similar taxes in place and public health officials in many other countries have been pushing for a tax on sugary drinks to reduce their consumption. It will be interesting to see if such measures do indeed reduce the market for sugary soft drinks and, in turn, have an effect on obesity rates.

What are artificial sweeteners?
With a decrease in sugar, this naturally leads to a greater use of, and subsequently focus on, artificial sweeteners. As the spotlight on the dangers of sugar continues, more consumers are turning to sugar substitutes to get their sweet fix. And it's not just consumer preference that is driving the rise in the popularity of artificial sweeteners. Some health professionals are recommending switching to sugar alternatives for those living with, or at risk of, obesity or diabetes, for instance, as part of a healthy diet and to aid weight loss.

Artificial sweeteners, also referred to as sugar substitutes, are calorie-free or low-calorie alternatives to sugar; simply food additives used to sweeten foods and drinks. A range of sweeteners is used in the manufacture of soft drinks and the most commonly used include: sucralose, acesulfame potassium (or acesulfame-K), aspartame, saccharin and steviol glycosides. Sweeteners are made from different raw materials and compositions that can result in a wide range of properties, from taste and the areas of the sweetness receptor they bind to, to their chemical stability. For instance, you may be trying to develop a particular taste profile with a certain combination of sweeteners, or trying to mask a bitter after-taste of one of the sweeteners. In general, they are many times sweeter than sugar, and so can be used in very small amounts, and most are excreted unchanged through the body. On this basis, they are considered metabolically inactive with no physiological effect on us. However, some researchers believe that despite this, they may still be interacting with the gut microbiome and could potentially upset our delicate microbial ecosystem and cause ill health. Sucralose, aspartame and saccharin have all been found to disturb the balance and diversity of the microbes in our gut. Numerous studies have investigated potential beneficial and adverse effects of sweeteners but the evidence is so far inconsistent. Some small studies have flagged up potential issues but when the findings are pooled across studies, no conclusive evidence of harm has as yet been demonstrated.

Sucralose is a calorie-free sweetener over six hundred times sweeter than sugar that is found is a wide range of low-calorie foods. Perversely, as it is so sweet, sucralose is often mixed with non-calorie-free sweetening ingredients (such as dextrose or maltodextrin) to dilute its sweetness. Sucralose is made by chlorinating sugar (in the form of sucrose – your typical table sugar). Some have linked sucralose with side-effects, such as migraines and problems with the immune system, but large-scale research has not confirmed this. The EU Scientific Committee on Food has determined that sucralose has acceptable safety for general consumption and it is approved for use by government agencies worldwide.

An ingredient of many low-calorie foods and drinks, acesulfame potassium is calorie free and around two hundred times sweeter than sugar. Although acesulfame potassium is considered safe for use by both the US FDA and the EFSA, not all are convinced. Some critics believe that there hasn't been enough good quality research to demonstrate its safety and that it may pose potential harm to our health, linking it to a number of diseases. There has been some recent work suggesting that long-term consumption of acesulfame potassium may alter gut bacteria and even promote weight gain, but further research is needed to confirm such a link.

Aspartame is an additive that has been particularly scrutinised. It is around two hundred times sweeter than sugar, with minimal bitterness, and used in thousands of food and drink products. When digested, aspartame is broken down into aspartic acid, phenylalanine and methanol. Aspartic acid and phenylalanine naturally occur in many foods that contain protein, and methanol is found in fruits and vegetables and their juices. While methanol can be toxic at high levels, the quantities stemming from the metabolism of aspartame are lower than those found in many other food sources. Aspartic acid and phenylalanine are considered safe for most of us, although a small number of people[3] need to be careful to avoid sources of phenylalanine, including aspartame, as their bodies cannot break it down. In the early 2000s, a number of studies by the European Ramazzini Foundation linked aspartame consumption with the increased risk of cancer, particularly lymphomas and leukaemias. However, these studies were in rats, not humans, and their research quality was criticised by the EFSA. The EFSA considered all the evidence available and reconfirmed the safety of aspartame. In 2006, the US National Cancer Institute carried out its own research into the effects of aspartame in people (285,079 men and 188,905 women) and found no evidence to support the hypothesis that

3 Those with the rare genetic disorder called phenylketonuria.

aspartame increases the risk of leukaemia, lymphoma or brain cancer. In 2013, the EFSA re-evaluated the safety of aspartame, based on previous evaluations, additional literature and new data, and concluded that it was not a safety concern at the estimates of consumption levels or at the Acceptable Daily Intake (ADI).

Saccharin, around three to four hundred times sweeter than sugar, has a slightly bitter and metallic aftertaste. It was discovered in the nineteenth century by chemists Ira Remsen and Constantin Fahlberg and is the oldest artificial sweetener. Despite concerns in the 1970s that saccharin may be associated with cancer risk, the European Scientific Committee for Food and International Agency for Research on Cancer, among others, have evaluated this sweetener and have concluded that there are no concerns about its safety profile.

Steviol glycosides are derived from the purified extracts of the stevia plant. They are positioned as a more natural alternative to sugar. Again, they are calorie-free sweeteners, somewhere between two hundred and three hundred times sweeter than sugar, used in many foods and beverages. The EFSA has deemed them safe on the whole but estimates that many of us consume more than the ADI, particularly from soft drinks.

With the rising consumption of artificial sweeteners, their increased and long-term use will face further scrutiny to ensure we're not replacing one so-called evil (sugar) with another (sweeteners). As artificial sweeteners are much sweeter than sugar, their burgeoning use may change taste preferences towards sweeter products and somewhat defeat the purpose of switching from sugar. People who frequently consume artificial sweeteners may start to find less intensely sweet foods, like fruit, unappealing and preferentially select very sweet and artificially flavoured foods and drinks over more nutritious alternatives. This may be particularly concerning in children if their tastes are conditioned when young and these taste preferences continue into adulthood.

> **Supertasters and sweeteners**
>
> Some people experience tastes more intensely than others. They are more sensitive to bitter flavours and fatty and sugary foods, and can have a particular dislike of broccoli and spinach, dark chocolate and coffee, for example. These people are referred to as supertasters and may have a particularly high number of taste buds on their tongue (although this is still being debated). Around one in four people is a supertaster, of which I am one[4] and this may explain why I've always been averse to artificially sweetened drinks. From as young as I can remember I could usually tell if a drink had artificial sweetener in it, and this same sensation has been found among other supertasters, whether sensitive to potential bitter aftertastes of sweeteners or the heightened sweetness of them.

What are flavourings?

Sweetness is one aspect of flavour but what about other flavourings? (Listen up . . . I'm going to reveal what's in Coca-Cola!) In comparison with other ingredients in our soft drinks, flavourings are used in relatively small amounts so our exposure to them is pretty low. Flavourings may be natural, deriving from fruits, vegetables, nuts, barks, leaves, herbs, spices and oils, or artificial (i.e. manufactured synthetically), deriving from compounds such as esters, aldehydes, alcohols and acids, and either impart a particular flavour to a drink, replace a flavour that's been lost during the manufacturing process, or enhance a flavour. Despite the distinction in name, in reality the lines between what is natural or not may be blurred. For example,

---

4 As confirmed by a special test strip. I also have a strong sense of smell, which is linked to how we taste. A curse at times!

the 'natural' flavour may not actually be the exact ingredient you think it is, e.g. raspberry, but could be a chemical originally found in raspberries that has been extracted and enhanced then added back to your drinks product. Natural flavourings are more expensive than those that are synthetically produced and are prone to fluctuations in weather, environment and civil unrest, whereas artificial flavours are stable, consistent, readily available and more cost effective. In general, flavours need to be water soluble so that they disperse evenly throughout the entire drink and they come in the form of extracts, i.e. flavouring oils dissolved in alcohol and water and often used in clear drinks like lemonade or ginger ale, or emulsions (blends) where the flavour oils are suspended in solution, equally distributed, and produce a cloudy drink.

As the amounts in your drink are small, manufacturers need only list the word 'flavourings' in the ingredients rather than the exact flavourings. This could refer to tens of different flavours that have been added and it is very difficult to find out what they are. It also means that it is hard to assess the safety profile or, indeed, potential health benefits of such drinks components. Interestingly, food flavourings are not considered as 'additives' in the EU and so are not assigned E numbers.[5] This is not to say that flavourings don't have to pass strict safety regulations; they do. The amount of flavourings in drinks is really very small in the grand scheme of things and, as such, they are unlikely to have a big impact on our health, whether positive or negative, but transparency of what is actually in our drinks could be improved.

What are acidity regulators?
Acidity regulators are food additives that are used to either change or control the acidity or alkalinity of drinks. In essence, they help

5 E numbers are assigned to food additives. It means that an additive has passed safety tests and is approved for use in the EU.

The secret formula behind Coca-Cola

The blend of ingredients used to make Coca-Cola has been famously kept secret for over a hundred years. We know that it contains carbonated water (for the fizz), sugar (for the sweetness – in the UK, in other countries it uses high fructose corn syrup instead), caramel colour (for colour, obviously), phosphoric acid (for tartness), caffeine (to add slight bitterness) and 'natural flavours', also known as 7X (this is where the secret lies). In 2013, an historian called Mark Pendergrast published a book in which he claimed to reveal the original formula for Coca-Cola, after he came across the recipe during his research into the Coca-Cola company. Pendergrast is not the only one to claim to have found the secret recipe and others have performed lab analyses of this cola's contents. However, Coca-Cola refutes Pendergrast's, and others', claims, insisting the only copy of their recipe is held in a secure vault in Atlanta. Whatever the truth, it seems that the 'flavours' comprise a mix of oils and extracts, including lime juice, orange oil, lemon oil, coriander oil, neroli oil and vanilla extract, along with nutmeg, cinnamon, coriander, citric acid, ascorbic acid, cochineal and alcohol. It also contains extract of coca leaf, which is the source for cocaine. Coca-Cola got its name back in 1886 from its two key 'medicinal' ingredients: the coca leaf (for the stimulant) and kola nut (for caffeine). In reality, there was only a very small amount of cocaine in the original version of Coca-Cola, and it was completely removed by 1929. Also, the caffeine in today's version of the product no longer comes from kola nuts.

balance the sweetness in soft drinks. This aids the processing, taste and safety of products. Regarding this last point, inadequate control of a product's pH can result in unwanted bacterial growth and be a potential health risk. There are many acidity regulators approved for use in our food and drink, including citric acid (probably the most commonly used in soft drinks), malic acid and phosphoric acid. While the first two in that list generally raise little concern about their effects on health (within regular levels, of course), phosphoric acid has been linked to a negative effect on bone health.

Phosphoric acid is a key ingredient in cola drinks, but not in other sodas, and research has found that cola consumption, but not other carbonated beverages, is associated with lower bone mineral density. Researchers have linked the effect on bone health to both the caffeine and phosphoric acid contained in colas but the effect seems to be more pronounced in the caffeinated, versus decaffeinated, versions of the drinks. Phosphoric acid is also thought to be a contributor to increasing the risk of other health problems, such as harm to the kidneys and cardiovascular system, but the evidence remains uncertain and speculative.

How do fizzy drinks get their fizz?
Carbon dioxide (CO_2) is a non-toxic, safe, virtually tasteless gas that is used to add the fizz to fizzy drinks, hence the term carbonated drinks. CO_2 dissolves in water under pressure where it forms carbonic acid. It is in fact carbonic acid, and not the bubbles, that causes a tingly sensation on your tongue. To make a drink fizzy, CO_2 is pumped into it under pressure and the container sealed. This keeps the pressure up inside the container but when it is opened again the pressure is released and dissolved gas bubbles escape. As the CO_2 escapes, the beverage starts to go flat. We all know that sensation. Both acidity regulators and carbonation play a role in helping preserve soft drinks for longer.

What are colours?

You can probably deduce that 'colours' on an ingredients list refers to additives that aim to alter or enhance the colour of a product to ensure it is aesthetically appealing. The three main categories of colourings are: natural colours, artificial colours and caramels. Natural colours derive from plants, fruits and vegetables and their increasing use, in contrast with a decrease in artificial colours, in both the EU and US, has been driven by consumer preference for more natural products. Artificial colourings have long been subject to stringent regulations and scrutiny, amid concerns of any possible health effects. (When I was at primary school, I once turned a lovely shade of beetroot when I became covered head-to-toe in a rash thought to be the result of a red food colouring in my school dinner.) Certain artificial colours have been linked to allergic reactions and the intensification of asthma symptoms, and possibly hyperactivity in children. Despite similar regulatory frameworks and safety assessments across parts of the world such as the EU and US, a number of colourings are permitted for use in some countries but banned in others, including ponceau 4R (E124), quinoline yellow (E104) and carmoisine (E122). In 2008, the UK's Food Standards Agency urged for an EU-wide ban on these particular colours, along with tartrazine (E102), sunset yellow (E110), and allura red (E129), amid safety fears. The EU rejected its calls and these colours are still permitted in our foods and drinks. The quality of the scientific evidence regarding potential harms has been questioned and there remains uncertainty as to the overall safety of these additives. However, it is worth noting that very few of the drinks we consume actually contain these colours nowadays.

What are preservatives?

Preservatives do the job their title suggests – they are added to soft drinks to: a) extend their shelf life by slowing or preventing the growth of micro-organisms, such as yeasts, bacteria and moulds, b) prevent the breakdown of vitamins and minerals, and c) prevent discolouration in drinks. Not all soft drinks contain them but those containing fruit juice tend to need preservatives to prevent microbiological spoilage. There are four main preservatives permitted for use in soft drinks: sorbates, benzoates, sulphites and dimethyl dicarbonate.

Sorbates are very effective against yeasts, moulds and bacteria but they can affect the taste of a product. They are often used in combination with benzoates, especially where the soft drink is highly acidic. The most common preservatives used in soft drinks are potassium sorbate and sodium benzoate. Potassium sorbate has not been found to be harmful to health and, incidentally, there are no safety concerns about dimethyl dicarbonate based on its current use.

Sodium benzoate has been widely used as a preservative for over a hundred years. In certain conditions, something called benzene can form in very small amounts in some carbonated drinks that contain sodium benzoate when ascorbic acid (vitamin C) is also added. This may be exacerbated during extended storage at elevated temperatures. Benzene is classified as a carcinogen (i.e. capable of causing cancer), prompting safety fears. However, over the years, processing techniques have been refined to minimise benzene formation and the use of benzoates has been reduced to prevent or minimise its occurrence. If found in products, it is at such low levels that it is not thought to pose a safety concern for consumers. Despite this, an influential study published in *The Lancet* in 2007 found a link between sodium benzoate and hyperactivity in children, leading public health campaigners to call for its ban in food and drinks. Another study, from 2010, showed symptoms related to Attention Deficit Hyperactivity Disorder in college students who drank drinks rich in sodium benzoate. It is not clear how the levels of sodium benzoate ingested in the studies compare with levels typically found in soft drinks today, but questions remain as to whether we should still be using this preservative in our drinks.

Sulphites have been used as preservatives for centuries. There are old references to their use in Roman times when sulphur was burnt and unfermented juice exposed to the smoke to help in wine preservation, along with other uses. Sulphites were also added to casks of fruit juice back in the nineteenth century to help preserve it. These days, sulphites are commonly used in soft drinks, as well as alcoholic drinks, such as beer, cider and wine, to stop them fermenting. Adverse health reactions to sulphites have been documented. Individuals with a particular sensitivity to sulphites have experienced rhinitis, skin reactions or stomach complaints, for instance, but the vast majority have been asthmatic symptoms. Consumption of soft drinks containing sulphites has precipitated respiratory attacks in asthma sufferers, and it seems to be asthmatics who are most likely to suffer a reaction from this preservative. It is not clear how sulphites cause these ill-health effects. It is known that sulphur dioxide, released by sulphites, is an irritant to our airways and a gas that contributes to air pollution and acid rain; however, it is thought that sulphites cause sensitivities via multiple mechanisms.

What are stabilisers?

If you ever read lists of ingredients on the things you consume, it's likely you'll have seen the word 'stabilisers'. These are additives that have a number of roles, such as combining flavours, suspending particles, stabilising protein or enhancing the mouthfeel of a product. Some stabilisers are important to the look and feel of a drink, whereas others are more focused on the process of manufacturing a suitable end product. Because the drinks industry is expected to create a wide range of beverages containing particular nutrients, minerals and other savoured additives, stabilisers are needed to ensure these added extras don't compromise each product's flavour, appearance or texture. There is a large number of stabilisers to choose from and their inclusion in a drink will depend on the other ingredients and the desired end product. Examples include guar and locust bean gum, pectin and xanthan, which help to improve the mouthfeel in diet drinks and reduce separation in fruit juice drinks, for example.

Now we've got through the technicalities, let's take a look at the main soft drinks themselves.

Still and juice drinks

After that long discussion of drinks ingredients you might think there is not much left to say about still and juice drinks. Surely they're mainly fruit juice and water, right? Intriguingly, no, there's more to them than that. The difference between still drinks, fruit juices and nectars is actually set out in legislation that defines each by how much fruit juice they contain. Products labelled as juice must contain 100 per cent fruit juice, with no sweeteners, preservatives or artificial colours. Fruit nectar actually refers to a juice (technically a purée) that is too thick to drink as it is (e.g. from apricots or pears) and so must be diluted with water, and sugar added, to make it drinkable. Such drinks contain between around 25 per cent and 99 per cent juice. Still drinks typically refer to non-carbonated (hence 'still' and not bubbly) drinks with less than 25 per cent fruit content.

What is in fruit juices and smoothies?

Starting the day with an OJ[6] is a morning routine for millions. The cool temperature quenches our thirst, the citrus zing is refreshing and invigorating and the sugary content gives us the energy boost to wake up and get going. Fruit juices, of which there are many varieties, are very popular. They satisfy our need to hydrate as well as stir that feeling that we're consuming a healthy alternative to sodas and other sugary soft drinks. Whether fresh or concentrated, fruit juice is just juice but what does that actually contain? Almost 90 per cent

6 Orange juice.

of orange juice and apple juice, for instance, is water and the bulk of the remainder is sugars, particularly glucose (about 2 per cent), fructose (2.4 per cent in orange juice, 5.5 per cent in apple juice) and sucrose (4.2 per cent in orange juice, 1.8 per cent in apple juice). In much tinier quantities, there is protein and micronutrients including vitamins C and E, B vitamins including folate, and minerals – potassium, chloride, calcium, magnesium, phosphorous, sodium, iron, copper, zinc, manganese and selenium. All those nutrients sound amazing, but let's circle back to the sugar content.

In 2014, a significant blow was dealt to the healthy reputation of fruit juice as a drink. Concerns about the sugar content of fruit juice had been rumbling for years but the issue hit the headlines when a primary school headteacher in East London imposed a new rule to confiscate fruit juice cartons from children's lunchboxes. In the same period, a government advisor and senior figure in nutrition, Professor Susan Jebb, lambasted the official advice that juice counts as one of our daily recommended five portions of fruit and veg. She highlighted that juice doesn't contain the same benefits as whole fruit and, because your body absorbs it very quickly, it may cause your blood sugar to rise rapidly too. Unlike juice, whole fruit contains fibre that slows down the absorption rate, and doesn't produce a spike in blood sugar levels. The fructose content of juice, without the benefit of fibre to help us feel full, may be causing us to eat more. Some research suggests that, while consuming glucose helps to tell the brain that we're full, fructose has little effect on the brain regions that curb appetite and may even prompt craving, potentially leading us to consume more calories than we need. The main advice from public health experts is to eat whole fruit instead of juice but if we do drink it, we should aim for small servings, diluted with water and preferably drunk with a meal to slow down its absorption rate.

Whilst we're on the subject of fruit juices, it's worth clearing up confusion about terms you might see. Fruit juices are frequently used as an ingredient in soft drinks but come in different forms. A fruit 'concentrate' is where water is removed from the fruit leaving a product that is many times more concentrated and smaller in volume; it has a longer shelf life than regular juice and is much easier to pack, ship and store. Reconstituted juice refers to taking the concentrate and adding water back into it. Sounds daft, I know, but it is economically preferable because you can transport your concentrate around to where you need it (i.e. to other factories) and then reconstitute it as required in the development of new drinks. 'Not from concentrate' juice should not be confused with fresh juice. 'Not from concentrate' refers to juice that is extracted from fruit in the country of origin, lightly pasteurised then frozen or chilled and transported to the country where it will be packed. 'Freshly squeezed' juice is extracted from the fruit and for immediate use. A fruit 'comminute' is a purée made from the whole citrus fruit, including the skin and pith, then sieved to remove unwanted components such as seeds. The inclusion of the skin and pith results in a different flavour from juices made without these. Most fruit juice is subject to mild pasteurisation before it is packaged to ensure a safe product. Freshly squeezed juice may be subject to little or no pasteurisation.

A good smoothie at the right time can be a very satisfying thing. Commercial smoothies generally contain a combination of crushed fruit, purées and fruit juice. And the amount of fruit in each bottle is a lot more than a regular juice. They may also include other ingredients such as yoghurt, milk, plant extracts or sweeteners, like maple syrup or honey. They can give us the impression that we are consuming our daily quota of recommended portions of fruit in one easy go. Many contain as much, if not more, sugar and calories than a cola drink (although the type of sugar, and the body's response to it, is not ncessarily equivalent). Those that include dairy and other non-fruit ingredients can have a much higher calorific content. For example, you won't be surprised to learn that the Vanilla Bean and Maple Syrup Smoothie from Marks & Spencer, in the UK, contains more than double the amount of calories in a classic Coca-Cola. (It is pretty delicious though!)

In 2013, researchers from Harvard University published their findings of an analysis of data gathered between 1984 and 2008 from 187,382 participants in three long-running studies. They found that while greater consumption of whole fruits, particularly blueberries, apples and grapes, was associated with a lower risk of type 2 diabetes, the opposite effect was found with drinking fruit juice. Greater consumption of fruit juice was actually associated with an increased risk of type 2 diabetes. Other research has demonstrated that eating whole fruit is more likely to make you feel full than drinking fruit juice or smoothies. So, the problem here is that across a day you're likely to consume more calories if you take your fruit in liquid form as it doesn't fill you up so you eat more of other foods to satistfy your hunger.

Another popular drink in some parts of the world, particularly the Indian subcontinent, is lassi. These drinks may be sweet and/or salty and are based on a blend of yoghurt or buttermilk with water and added ingredients such as puréed fruit, sugar or spices. Mango lassi is a popular choice. In essence, lassi is a kind of smoothie. Many claims about the health benefits of lassi are made, not least because it contains yoghurt (generally good for the gut) and fruit. How healthy these drinks are will very much depend on the combination of ingredients. Some may contain a lot of sugar, if sweetened with honey, or they contain a lot of fruit purée, and some may be relatively high in fat if made with coconut milk and full-fat yoghurt, for instance. We've already discovered that liquid foods are not as filling as solid ones so consumers may need to be careful not to take in too many calories from a lassi drink on top of their usual solid food intake.

Dilutables: squashes and cordials
Squashes, also known as cordials, are concentrated syrups, typically fruit-flavoured (e.g. orange, blackcurrant, lime) that we add water to for a refreshing cold (or hot – a lot of people in the UK

grew up on hot blackcurrant squash) drink. In the UK, the most popular flavour is orange but the market has broadened over the years as consumers seek a wider choice of ingredients, prompting the emergence of flavours such as raspberry, elderflower and pomegranate. Dilutable drinks are very popular as they offer low-cost, easy-to-use products that can safely sit in your cupboard on standby. The vast majority are now low- or no-calorie versions. These drinks usually contain water, concentrated fruit juice, acidity regulators, sweeteners (some drinks include sugar), preservatives, stabilisers, thickeners, colours and flavourings. Barley waters also include barley flour and, as a result, must be labelled as containing gluten. For those of you familiar with Vimto and have ever wondered what is in it, let me enlighten you. As well as the ingredients I've already mentioned, the Vimto flavouring is listed as comprising natural extracts of fruits, herbs, barley malt[7] and spices. All right, I appreciate that this isn't very specific but as we discovered earlier, the manufacturers don't have to provide a detailed list of the flavourings they use. This is why so many drinks can claim to have secret blends of flavours.

Flavoured water

What we're talking about here are commercially made drinks that are marketed as water with a flavour added, as opposed to soft drinks in general or ones you might make at home. Flavoured waters are typically comprised of over 99 per cent water (be it spring or mineral, still or carbonated) plus flavourings, preservatives and sweeteners. The flavourings often comprise a tiny amount of fruit juice. Most often, the preservatives used in flavoured water are potassium sorbate and dimethyl dicarbonate, which are probably nothing to worry

---

7 Despite the inclusion of barley malt, the amount of gluten it contains is so low that it can be classified as gluten free. (Vimto, 'FAQ'. http://www.vimto.co.uk/faq.aspx)

The tennis players' favourite drink?

Robinsons Barley Water is a well-known fruit cordial in the UK, where many children grew up drinking it. Its brand profile is particularly strong due to its long-held partnership with the Wimbledon tennis tournament, its bottles of barley water perennially sitting under the umpire's chair for players to use. Of course, these days, tennis players have their own carefully balanced hydration drinks on court and don't touch the barley water. Many haven't even noticed it and those who actually try it off-court tend not to be that keen (it seems that it's more to the kids' taste). The association with Wimbledon goes back all the way to the 1930s when lemon barley water was created to hydrate players at the tournament. It then went into commercial production and the partnership between the drink and the championship continues to this day. In fact, in 2015 it extended its partnership agreement to 2020, making it the second-longest sports partnership in history.

about. While some flavoured waters contain sugar, the majority are marketed as sugar free and instead contain sweeteners, typically sucralose, acesulfame potassium and steviol glycosides. You often find that more than one sweetener is included. Flavoured sparkling water drinks have been shown to be potentially erosive on teeth and professionals suggest that they should be considered as acidic drinks rather than just water with added flavour (a different mindset altogether).

Other still soft drinks include iced teas and coffees, enriched waters and dairy or dairy-alternative based drinks. These are covered elsewhere in this book.

Carbonated soft drinks

Classic fizzy pop is still a consumer favourite. Carbonated soft drinks, sometimes referred to as sodas, make up the largest category of soft drinks in the UK, with a 38 per cent market share in 2016. There's a similar picture elsewhere. Fizzy drinks have been around since the late eighteenth century when the carbonation of soft drinks was invented by the English chemist Dr Joseph Priestly. There's a plethora on the market from lemonades and colas, to fizzy fruit drinks and flavoured waters, to tonic waters and mixers.

The process for manufacturing carbonated drinks is relatively straightforward but involves technical know-how and careful blends of ingredients. The water at the base of the drink needs to be free from potential contaminants to start with so a number of sterilisation processes, including boiling, filtering and chlorination, take place to make that happen. The main ingredients are then added to the water, either as powder or syrups, and the whole lot sterilised again in case of any contaminants present in the added ingredients. The fizz is then added and the resulting product bottled, ready for labelling and packing.

In terms of our health, it is pretty much a no-brainer to appreciate that sweet fizzy drinks aren't great. The previous sections on sugar, sweeteners and other soft drink ingredients cover the main problems associated with these drinks.

What are functional drinks?

The term 'functional drinks' refers to those beverages that aim to offer added health benefits. They are often developed to provide (in theory) specific health benefits, such as boosting energy and athletic performance, helping digestion, improving nutritional intake and promoting heart health. This category includes sports drinks, energy drinks and wellness drinks. Let's consider these in turn.

Scotland's other national drink

SCOTTISH READERS LOOK AWAY NOW! . . . Irn-Bru is often referred to as Scotland's other national drink (whisky, of course, holds the true title) and the Scots certainly do love it. So much so that once it was announced that the manufacturers were going to slash the amount of sugar in the (very highly sugared) recipe, some Scots began to stockpile the stuff. Irn-Bru is a bright orange, very sweet, carbonated drink famously brewed to a secret recipe of thirty-two flavours. Despite it being a symbol of Scottishness, the origin of this drink actually lies elsewhere. The original essence labelled as IRONBREW was created by the Maas & Waldstein chemicals company of New York in the late nineteenth century. The flavoured pre-prepared syrup was also imported and used by companies elsewhere to produce drinks with their own Iron Brew label. Over time, local producers developed their own recipes and Iron Brew became a generic term to refer to the type of drink, in the way we use the terms cola or lemonade. The producer of today's Irn-Bru, AG Barr & Co., launched their own original recipe in 1901 and featured a famous highland athlete, Adam Brown, on its label. And so the link between Iron Brew and Scotland was born and since firmly defended. Much has changed over the decades, not least its name to become Irn-Bru in 1947 following strict new labelling regulations after the war, but the Scottish love affair with the Day-Glo beverage continues. It is the best-selling soft drink in Scotland. Despite its name and jokingly being advertised as 'made from girders', it only contains a miniscule amount of iron.

What are sports drinks?

The idea underlying sports drinks is to help athletes hydrate effectively before, during or after exercise. They may be still or carbonated, ready-to-drink or in the form of soluble powders or concentrates, and fruit or non-fruit flavours. As well as water, they contain a lot of sugar for energy and mineral electrolytes, such as sodium, potassium and chloride. There are three main types: hypotonic drinks, isotonic drinks and hypertonic drinks that claim to have different functions. Hypotonic drinks provide a lower concentration of salts and sugar than the human body, isotonic drinks contain the same concentration of salts and sugar as the body, and hypertonic drinks contain a higher concentration of salts and sugar than the body. Both hypotonic and isotonic drinks are designed to replace lost fluids rapidly during exercise but isotonic drinks have additional sugar for an energy boost. Hypertonic drinks are best drunk after exercise and are designed to supplement an athlete's sugar intake to provide maximum energy (most suitable for endurance events where energy needs are particularly high) and are more slowly absorbed than other energy drinks. There are very few hypertonic drinks on the market and most sports drinks are isotonic in nature. That's all very convincing, but do they really work?

A randomised controlled trial of tennis players found that sports drinks attenuated players' fatigue and also made the players feel like they were using less effort during play. However, this small study was carried out in conjunction with a sports nutrition company and involved two of their employees, so these might not be the most objective results. In fact, a large number of studies into the effectiveness of sports drinks on athletic performance have been carried out, or sponsored by, companies with a vested interest in the products. Despite decades of research, little convincing evidence has been published to prove definitively the worth of sports drinks. The studies often only involve small numbers of participants, and the participants are often not representative of the people who frequently consume the drinks. The *British Medical Journal (BMJ)* was so concerned about the claims of the sports drinks industry, the links between the industry and research and potential negative effects of the drinks on consumers that they launched their own investigation. In July 2012, the *BMJ* published a series of articles looking into the sports drinks industry and how evidence is assessed by regulatory bodies. They found that sports medicine research is hugely entwined with the sports drinks industry, with many institutes and academic journals having long relationships with manufacturers. It is implied that the available evidence on the credibility of sports drinks cannot be viewed objectively and what does exist does not robustly and convincingly demonstrate their effectiveness, particularly not in the general population to whom such drinks are aggressively marketed.

There's significant concern about the potential to promote weight gain as many consumers ingest the surplus calories in the drinks without sufficiently expending enough energy to compensate for them. Regardless of the lack of evidence for their efficacy, sports drinks are supposedly specifically intended for individuals who are performing high-intensity exercise over an extended period of time (e.g. one hour or more), yet a survey in 2012 found that there may be as many as one in four adults in the UK who are consuming sports drinks while just sitting at their desks. So while you may think you're healthily hydrating, sipping sports drinks while walking round the park clad in lycra or after doing a few leisurely lengths in the pool is likely only to lead to weight gain. There is also concern from dental specialists about potential harm of sugary sports drinks on consumers' teeth. Unfortunately, sports drinks are particularly attractive to teenagers and many of this population group believe them to be a healthy drink. A 2014 survey found that only 16 per cent of UK teenagers who drink sports drinks do so for their designed purpose. Another study, in 2016, found that almost 90 per cent of Welsh twelve- to fourteen-year-olds drank sports drinks. At a time when child and teenage obesity rates are soaring, the consumption of additional, needless calories should be a significant concern. And it is. A number of studies are looking into links between sports drinks consumption and weight gain. For example, one prospective cohort study of over seven-and-a-half thousand children and adolescents in the US, between 2004 and 2011, found that the frequency of drinking sports drinks predicted greater increases in their body mass index (BMI[8]). In other words, there appeared to be a link between the sports drinks and weight gains in the participants.

What are energy drinks?
Another big health talking point, particularly among the parents of teenagers, is energy drinks. Energy drinks now comprise a huge part of the soft drinks market. In the UK, it is estimated that the energy drinks market is worth over £2 billion, and in the US nearly $10 billion. In the US, men between the ages of eighteen and thirty-four consume the most energy drinks, and almost a third of teenagers drink them regularly. UK teenagers reportedly consume the most energy drinks in comparison with their peers from sixteen other European countries.[9]

Energy drinks offer the consumer an energy boost, mainly provided by sugar and caffeine, as well as other stimulants. As a comparison, the amount of caffeine in energy drinks varies but in a 250ml-sized can is around the same as an average mug of coffee. These drinks are often sold in cans double that size, however, so the caffeine, along with other constituents, is consumed in twice the amount. Guarana is often used as an ingredient in energy drinks. It hails from the guarana plant, *Paullinia cupana*, and is traditionally consumed by indigenous communities of the Amazon region. It is prized for its stimulant properties, due to its high caffeine content,

8 BMI uses a person's weight and height to calculate whether their weight is within a healthy range.
9 At 3.1 litres per month versus an average of 2 litres per month.

hence its use in energy drinks. The seeds of the guarana plant are the size of coffee beans but they contain around twice the caffeine. Other caffeine-rich plant extracts used in energy drinks include tea, ginseng and yerba mate. Energy drinks have a high sugar content, although a few sugar-free varieties are also available, and also contain ingredients such as herbs, minerals and vitamins. They are mostly carbonated and contain taurine and glucuronolactone. Taurine is an amino acid that your body can manufacture itself that plays a role in cardiovascular, muscular and nervous system functions, amongst others, and is said to impart energy to the drinker. Glucuronolactone, found naturally in the body, is produced by the metabolism of glucose in the liver and is included in energy drinks to, purportedly, fight fatigue and promote a sense of wellbeing.

Should we worry about energy drinks?
A 2017 review of the evidence behind the effects of energy drinks concluded that, on the whole, their negative effects outweigh any positive effects. Some studies report a short-term, temporary improvement in mental and physical stamina among both adults and adolescents, as well as improved perceived alertness and restoration of fatigue. However, the majority of studies, of both short- and long-term consumption of energy drinks, suggest negative health effects, particularly attributable to caffeine and sugar but also highlighting the need for more research into the effects of the other ingredients. The benefits or potential harms of taurine and glucuronolactone, for example, are unclear as they have been relatively little studied. As even the British Soft Drinks Association notes, there is no evidence that taurine has any beneficial effect to a healthy person. A range of side-effects from drinking energy drinks has been commonly reported, including headaches, increased blood pressure, irritability, sleep problems and stomach aches. These are thought to be largely due to the caffeine content of the drinks. There is also emerging evidence of an association between energy drinks

and problematic risk-seeking behaviours, cardiac issues, kidney and dental problems. Hyperactivity is also commonly reported and there has been a significant number of injuries and even deaths blamed on over-consumption of energy drinks. A particular problem is that many people combine energy drinks with alcohol and this may make them feel less intoxicated than they actually are, prompting more risk-taking behaviours. The sugar content is also a big issue, with reports that energy drinks may contain 60 per cent more calories and 65 per cent more sugar than other soft drinks. Just as with sports drinks, this means people are consuming far too many unnecessary calories in their beverages and energy drinks consumption may also be fuelling weight gain.

Concerns have arisen about what effects energy drinks may be having in children and adolescents. Teenagers and young adults are the target market for these drinks, but little specific research into the effects of such beverages on this population group has been carried out. Young people may be more vulnerable to the effects of energy drinks, particularly the high caffeine content, than adults as their bodies are still developing and they weigh less so they experience greater exposure to the stimulant ingredients. The high sugar content is also undesirable and may affect their taste, in the longer term developing a preference for highly sweetened foods and drinks. There is a strong link between energy drink consumption and alcohol use, as well as high-risk behaviours. Even without these drinks, teenage years and young adulthood are known to be times of experimentation and risk-taking for many as they develop their independence, so throw energy drinks and, potentially, alcohol into the mix and negative actions and consequences may be intensified. A number of studies have found serious adverse outcomes among college students who take energy drinks and alcohol together, including a higher risk of being assaulted or of assaulting someone, being injured, and requiring hospital treatment. Frequent consumption of energy drinks, especially when mixed with alcohol, may also be a marker of an increased risk of the substance use. Young people combining high-risk behaviours with health issues like cardiac problems, eating disorders or anxiety may be particularly vulnerable to the effects of energy drinks. Although some young people may drink energy drinks in the belief that they help with their concentration for studying, research has found that too much caffeine can actually harm your cognitive abilities. While students on energy drinks felt that they were more alert and stimulated during research tests, they performed less well than their peers who did not consume the energy drinks.

Such is the concern about the effect of energy drinks on our health, warnings about them have been issued by a number of health bodies, and several European countries, such as Sweden and Lithuania (and soon the UK), have even banned their sales to children. A number of countries have also placed a high tax on energy drinks in an effort to reduce their popularity.

Military powered by caffeine

Research suggests that US military forces consume substantially greater amounts of caffeine than age-matched civilians. The source of that caffeine varies across generations, with younger cohorts favouring energy drinks over coffee. This preference for caffeine is seen in both deployed and non-deployed personnel.

What are wellness drinks?

Now we're opening up a whole can of worms. There has been an explosion of so-called wellness drinks in recent years, driven by rising consumer interest in health and wellbeing and a desire for added extras in their beverages. Who wouldn't want to look beautiful and feel amazing with just the effort of drinking an attractively packaged beverage? The world of online bloggers and health and beauty websites have jumped on this trend and catapulted wellness products into the mainstream, jumbling up science jargon, marketing hype and Instagram-able images for the greatest influence.

'Wellness drinks' isn't a technical term, as such, more of a catch-all to describe drinks that aim to promote wellness in one way or another, whether by providing additional vitamins and minerals or other health-boosting nutrients, for instance. Of course, if you were being pedantic you could say that water is the ultimate wellness drink, or that removing or lowering the sugar content of drinks is a move towards the better health worth of products. Juices and smoothies are also a go-to for those trying to choose a healthier alternative to a sugary fizzy drink, but here I'm particularly exploring those beverages that specifically target the wellness market with claims about their health-affirming contents, such as those that aim to offer cognitive, immune system or beauty benefits.

Plant-based drinks

The assumed power of plants and their inherent valuable nutrients are the basis of many wellness drinks. Here, I'll focus more on the juice-type plant drinks, as dairy alternatives, like oat, nut and rice drinks, which are obviously also plant-based drinks, are covered in Chapter 2, Milks.

What is in aloe vera juice?

The aloe vera plant is a spiky cactus-like succulent, with thick fleshy leaves, typically found in tropical climates (and many people's kitchens). Aloe vera has been used for medicinal purposes for thousands of years, in places such as Greece, India, Mexico and China. Nefertiti and Cleopatra reportedly favoured it in their beauty regimens and Christopher Columbus used it to treat soldiers' wounds. These days, aloe vera is widely used in pharmaceutical, cosmetic and food products, including aloe vera juice drinks. You can purchase pure aloe vera juice but aloe vera juice drinks typically contain aloe vera juice along with water, sugar or sweetener, other fruit juices, flavourings, acidity regulators and stabilisers.

The aloe leaf gel contains over 98 per cent water and it is the other 2 per cent that contains potentially beneficial molecules. Over seventy-five potentially active components are found in aloe vera, including sugars, enzymes, vitamins, minerals and amino acids. The vitamins include antioxidant vitamins A, C and E, as well as vitamin B1 (thiamin), niacin, vitamin B2 (riboflavin), choline and folic acid. Potassium is reported to be in higher concentration in aloe vera juice than most plant products, and calcium, magnesium, copper, zinc, chromium and iron have also been identified.

Aloe vera gel has been found, in humans, to enhance the bioavailability of vitamins C and E when ingested at the same time, which could be useful for increasing the benefits of what we already consume. It is also hypothesised that aloe vera may aid the absorption and bioavailability of medicines, meaning that they become more effective and potentially lower doses may be used. Work is ongoing to investigate whether aloe vera might be helpful in this respect. Having said this, this is unlikely to refer to the regular aloe vera soft drinks that are readily available on the high street. These drinks only contain a small amount of aloe vera juice, watered down and sweetened. What you're mainly likely to gain from such beverages is unnecessary sugar.

Despite many claims regarding aloe vera's anti-inflammatory, antioxidant, anti-viral, anti-fungal, anti-diabetic and anti-cancer capabilities, to name but a few, very inconsistent research results have been produced, and solid intervention trials are needed to convincingly prove measurable benefits from aloe vera juice drinks themselves.

Juice from the prickly pear cactus is also now widely available as a drink. Similar to the aloe vera drinks, it may be purchased already added to water, other juices, sugar and flavourings. It is promoted as containing much less sugar than coconut water and being just as refreshing.

What is in coconut water?
Although it has long been drunk in the tropics, coconut water (not to be confused with coconut milk) has started being widely found in our shops in the last few years. Coconut water is the clear liquid from the inside of young coconuts. There is more water in the immature coconuts, and it is easier to get at than in older coconuts as the shell is softer. Commercialising coconut water on a large scale started off as an idea for reducing waste. The coconut kernel was already being exploited to produce milk, cream and desiccated versions, for culinary purposes, and oils for use in cosmetics and soaps, and the bran was being used as animal food and the husk repurposed in furniture. That just left the water. While collecting coconut water was a bit of a faff for the manufacturers initially, promoting it as a nutritious and refreshing beverage paid off. The coconut water market has exploded in recent years and market forecasters believe its popularity is likely to continue.

Coconut water is touted as a refreshing drink with many health benefits but is there any science to back this up? Fresh coconut water contains around 2.6g of sugar per 100ml but packaged coconut water that you buy as a drink tends to contain more, somewhere between 3.5g and 5g. You can buy large cartons of coconut water for serving your own portion but many bottles contain the same amount as a regular can of soda (330ml[10]). This product size of coconut water contains somewhere between three and four teaspoons of sugar;[11] that compares to around three teaspoons of sugar in a regular Sprite, nearly nine teaspoons in a regular Coca-Cola, or just over eight teaspoons in fresh orange juice

10 This assumes you drink the whole 330ml as one serving. Some coconut water brands suggest smaller serving sizes.
11 Based on 4g in a teaspoon.

(although the types of sugars in these drinks are not necessarily equivalent). So, on balance, coconut water has less sugar than many other soft drinks but what else does it contain?

Coconut water is low in fat (but then so are other soft drinks) but it does contain a wide range of vitamins and minerals. A 330ml bottle of coconut water contains over 670mg of potassium which is around 15–20 per cent of our recommended daily intake. This is one reason that some choose coconut water as a sports drink as athletes can lose a lot of potassium through sweating when undertaking intense exercise. Potassium is an essential mineral that plays a vital role in a range of bodily processes, including heart function, muscle contractions and maintaining fluid balance. (The high potassium content of bananas is also one of the reasons why sportspeople are often spotted chomping them. For reference, a medium-sized banana contains around 400mg of potassium.) Research has demonstrated that coconut water is just as good as sports drinks at rehydrating athletes after exercise, and for some it is easier to consume large amounts during rehydration, although it seems that the taste of sports drinks is preferred. There is a long list of other health claims associated with coconut water (e.g. aiding weight loss, improving sports performance, lowering blood pressure or cholesterol) but, as yet, the scientific data is not there to support them. To get the healthiest version of coconut water, make sure you read the label and avoid those with added sugar or flavours.

What are sap waters?
Ever heard of birch water? I hadn't until recently. It is another trend in the soft drinks market. Birch water is simply the sap tapped from birch trees. Long drunk in Baltic countries, Canada and China, among others, it is said to be clear, crisp and refreshing with an 'aftertaste of forest'.[12] Unlike most other soft drinks, it comes from

12 According to the manufacturer of Sibberi Pure Birch Water.

assorted small-scale harvests as the growing and harvesting of this product is apparently awkward and unlikely to yield high profits. A natural alternative to water and coconut water, birch water was labelled the new 'superdrink' for its supposed health qualities. But is it all that? Birch water contains much less sugar than coconut water, which is a bonus, and is promoted as being a good source of manganese. Manganese is involved in bone formation, energy metabolism and helping prevent cell damage. Commercially available bottles of birch water tend to be around 250ml in size, and contain approximately 0.3mg of manganese. Adequate intake of manganese is around 2mg per day but manganese deficiency is rarely seen, so highlighting the benefits of it as a source of manganese may seem rather futile. And beyond the manganese highlight, it seems there's little else to justify the high price of this refreshment. Water will probably serve you just as well. So, if you're going to buy it, buy it because you enjoy the taste not because you're expecting subsequent health benefits, and, once again, avoid the flavoured versions.

Birch water isn't the only tree sap being marketed as a drink that is gaining popularity. Maple water, not to be confused with maple syrup, is very much along the lines of birch water, i.e. the natural sap of the maple tree. (In case you're confused, maple syrup is made by boiling down the maple sap; it doesn't come straight out of the tree like that.) Again, it is lower in sugar than coconut water and actually contains more manganese than birch water.

Bamboo water is, unsurprisingly, the sap from the bamboo tree, and its taste is described as crisp, with 'hints of green tea and a smoky after taste'.[13] It is said to contain no sugar but is high in silica. Manufacturers claim this silica stimulates collagen production and, in turn, gives skin a healthy glow and makes hair shiny. Hmmm. That's all very well but are there intervention trials that actually prove this? In a word, nope. First, we need to know what form the silica is

13 According to the manufacturer of Sibberi Bamboo Water Glow.

in – is it the form that is the most bioavailable? Is it in high enough quantities to have an effect on our body? Have significant, measurable improvements in hair and skin been demonstrated in a lot of people in controlled experiments? Not yet.

There is no hard scientific evidence backing up any particular health effects of these sap water drinks. Some are pure products, in other words containing just the sap, but others have been diluted with water and flavoured with additional ingredients, which aren't ideal. Also, the unique combinations of vitamins, minerals and other nutrients of these saps are often marketed as being highly advantageous for our health but, as nutrition experts point out, humans and trees have very different nutritional needs, so this is nothing more than ridiculous marketing hype. Finally, they are all promoted as being 'hydrating' – maybe they are, but water itself is very hydrating and doesn't attract the high price tag.

What is kombucha?
You may have heard of it, but have you actually tasted it? I have and, for me, once was enough. Kombucha is a fermented drink comprising tea with sugar that has been exposed to a layer of culture called 'scoby' – symbiotic culture of bacteria and yeasts. Scoby is a beige, rubbery, alien-looking mass teeming with microscopic life. The sweetened tea is actually helping the scoby to grow; it is literally feeding a living mass. While preparation methods vary, kombucha is produced by adding scoby to a sweetened tea blend and leaving it to ferment for around one to three weeks at room temperature. The tea is then bottled with some extra sugar and left to evolve for a few more days. During this time, natural carbonation takes place giving a slight fizziness to the resulting brew. After this, additional flavours might be added and then the whole lot refrigerated to inhibit further fermentation, and it's good to go. So what does it taste like? Based on my own experience, I would describe the taste as slightly sweet, sour and vinegary, with an underlying effervescence that reminded

me of fruit juice that's long past its best.[14] The bacteria and yeasts in the scoby convert the sugar into ethanol (a form of alcohol) and acetic acid. This acid is what gives kombucha its tangy flavour. It is a little cloudy in appearance and also has bits floating in it; they don't feel nice in the mouth. 'Mmm, yummy' you're probably think-ing by now ... but if not, perhaps you're wondering why kombucha has quite the cult following. In fact, that term is doing a disservice to its popularity – in 2016, kombucha sales in the US were esti-mated at around $600 million and the mighty PepsiCo bought out a small kombucha company (*Kevita*) to get in on this growing market.

There is growing interest in the medicinal and biological value of kombucha. Proponents of this 'living' drink assert that it can do everything from improving digestion and promoting weight loss to controlling blood pressure and preventing cancer. But how much truth is there in all of this? Just like other fermented foods, such as yoghurt and sauerkraut, kombucha contains live micro-organisms. It contains probiotic bacteria, which can help balance our gut's microbiome, promote a healthy immune system and aid digestion. A number of animal studies have found that kombucha contains a variety of bioactive compounds that have the potential to support our health. However, we don't yet know whether kombucha drinks actu-ally contain enough beneficial bacteria and other bioactive compounds to be effective. Furthermore, it is not known whether such bacteria and compounds can survive the very acidic environ-ment of our stomach to ultimately make it into the gut to confer health benefits, or are in the right format to do so. The tea that kombucha is made from may itself confer some benefits. Tea is rich in polyphenols that have antioxidant properties and this component

---

14 In the interests of disclosure, I should point out that I intensely dislike vinegar anyway so am probably not the best person to take the word of on the taste of kombucha.

of kombucha may be good for our health. This wellness drink also contains some vitamins and minerals, such as B vitamins and vitamin C, but they are in tiny amounts and are unlikely to be noticed by your body once inside. But don't forget what else is present in kombucha, namely some sugar, caffeine and alcohol. Yes, these are in small amounts but it is too easy to dismiss any harmful effects of these ingredients because of their trivial quantity, while also believing in the potential beneficial effects of vitamins, minerals and other components despite them also being present in tiny amounts. You can't have it both ways!

When it comes down to it, there is practically no robust evidence to demonstrate the clinical effectiveness of kombucha. However, studies have shown some risks from drinking it. There are documented cases of individuals suffering serious adverse health effects after drinking kombucha, such as liver and gastrointestinal problems and lactic acidosis.[15] Because of potential risks and that not enough is known about the effects of kombucha on our health, certain groups of people are advised to avoid drinking it, including pregnant or breastfeeding women and those who are immune-compromised. It's fair to say that we just don't yet know enough about kombucha and how, or even if, kombucha is making its mark on our health.

Active ingredients
Other than promoting the natural-ness and/or organic-ness of their ingredients, wellness drinks often include additional ingredients to enhance their benefits to our health. They may sound credible and essential for living a long and healthy life (indeed, that's what makes them sell so well), but is there good evidence that they are making any substantial contribution?

---

15 Lactic acidosis is where excess acid builds up in the body. While it can usually be treated, if left untreated it may be life-threatening.

Extra vitamins are often added to soft drinks to increase their health potential. I've previously talked about the some of the issues surrounding these added vitamins, in Chapter 1, Waters, but needless to say that just because they were added at the factory, it doesn't mean they will make it into your system in a high enough quantity and useful form to do anything beneficial. B vitamins are commonly included in energy drinks and other functional drinks and the impression perpetuated is that these particular vitamins help to give you an energy boost. B vitamins are responsible for a wide range of bodily functions, from keeping your nerves and muscles healthy, and aiding the nervous and digestive systems, to promoting skin and eye health. They also help convert energy from other nutrients into the form needed by the body, but what they don't do is impart energy directly. Taking lots of B vitamins won't give you a quick pick-me-up and, in fact, you are probably already getting enough B vitamins to do the physiological work required; any extra amount will likely just be peed out. The addition of B vitamins in wellness drinks is a marketing ruse and any effects on your energy after drinking them will probably come from the sugar or any stimulant ingredients that they also contain.

Botanical extracts of fruits, vegetables, herbs and spices are popular themes within wellness drinks. The idea that the flavours are being provided by natural ingredients rather than artificial ones is appealing to consumers and their presence already evokes a feeling that you're drinking something healthy. Add to that some claims about the wholesome or medicinal value of such ingredients and you've got yourself a winning product. Various plants have been found to have antioxidant, anti-inflammatory or anti-cancer properties, for instance, although this is usually in other species and in much higher doses. Take turmeric, for example. This deep yellow Asian spice, more commonly associated with your Saturday night takeaway curry, has been a popular added extra in a wide range of drinks in recent years. By adding it to regular drinks, manufacturers have appealed to a broader market for their products by capturing the attention of those wanting more from what they consume. Turmeric is lauded for its wealth of health benefits. This is due to the curcumin that it contains. Curcumin, in various forms, has been found in clinical trials to have positive effects against many different diseases and has demonstrated anti-inflammatory and antioxidant properties. This is great, but the problem is that the doses used in the trials are far higher than you can get from regular turmeric powder that is used in drinks. And clinical trials on turmeric itself have not been so definitive. Only about 3 per cent of turmeric powder is curcumin, so you're not getting very much of it with just a sprinkle being added to most drinks. Even if you add more to your drink than most, there is another problem you'll encounter. Curcumin by itself is not very bioavailable. It is poorly absorbed, quickly metabolised and rapidly eliminated from the body, so you're not going to get much, if any, benefit from what you do consume. Where this compound has been found to be effective is when it is being used in combination with another agent, that improves curcumin's bioavailability. So in the end you would probably have to consume a huge amount of turmeric-enriched drinks, along with another agent to aid the curcumin's bioavailability, every day to gain any effect at all from the curcumin content. And you wouldn't want to drink too much as this has led to nausea and diarrhoea in some.

The Malaysian herbal drink, air mata kucing, is purported to be nutritious as well as refreshing. Containing monk fruit, longan, winter melon, water and sugar, this drink is served with lots of ice and is particularly popular when the weather is hot. As well as being a calorie-free natural sweetener, many times sweeter than sugar, monk fruit is used widely in traditional Chinese medicine as is thought to have anti-inflammatory properties. It is often brewed into hot drinks to help with the treatment of the common cold. Longan, a fruit similar to a lychee, is reported to be high in vitamin C and antioxidants. So the combination of ingredients is considered to deliver health benefits.

Chilled persimmon punch, known as sujeonggwa, is another well-known drink used to strengthen the stomach and prevent the common cold. This Korean drink is made with water, sugar, ginger, cinnamon, dried persimmons, walnuts and pine nuts. There is a lack of conventional scientific research demonstrating any measurable benefits of either of these drinks. These drinks are largely consumed due to traditional beliefs and anecdotal evidence.

You can't have failed to notice that many of the products we consume are labelled as containing either probiotics or prebiotics. There has been a huge surge of public and research interest in what's going on at the microscopic level in our guts. Apparently, we're all bloating, gassy and intolerant to everything and it's our guts that are to blame. Our guts are filled with microscopic creatures (mostly bacteria), around 100 trillion of them in fact, just trying to do their job, but they are ostensibly being upset by our terrible diets and lifestyles and need help to rebalance themselves. We want to keep them happy as they play a vital role, from helping metabolise nutrients from food and supporting our immune system, to producing vitamins and playing a role in defence against disease. But just what are probiotics and prebiotics, and do we really need them? Probiotics are living bacteria and yeasts that are often added to yoghurt drinks. They also occur naturally in kefir (see Chapter 2, Milks) and foods like yoghurt and fermented cabbage (such as sauerkraut or kimchi). Probiotics are thought to help restore the balance of your gut bacteria when it has been disrupted, for example through ill health or use of a particular medication like antibiotics. Prebiotics are non-digestible plant fibres that nourish the friendly bacteria in your gut. They are found in many plants and vegetables. It is true that we need to try and maintain a healthy gut as evidence is emerging all the time about the importance of the gastrointestinal microbiome for so many aspects of our health. However, it is not clear whether ingesting prebiotics and probiotics for this purpose is effective.

There are a number of questions over the use of probiotics. There are hundreds of different species of gut microbes, and products vary in the strains of bacteria and yeasts they contain, so are we getting the right ones? The thing is, we don't yet really know which ones we need. As individuals, we may need different types at different times. We're also not sure how many different types we need or even how much of each, so we don't know whether the products contain the right amount to make a difference to us. If we knew whether we were lacking a particular type and how much we needed to replace it, that could offer a promising route to addressing the balance. However, even if we knew those things, it is not clear whether all the probiotics we consume necessarily survive the acid environment of our stomach and so reach the gut to populate it. Some research has found that while a small proportion do survive,[16] to have the maximum effect on our health they need to stick in our intestinal cells and there isn't much evidence to suggest this is necessarily happening. As such, they aren't multiplying when they reach our guts and are passing through into our faeces. As they aren't sticking and multiplying, this means that you continually need to consume them to gain benefit. The effects are transient, which is counter to the assumption that you are somehow boosting your colony of microbes. There's even some evidence to indicate that the probiotics we introduce may have a negative effect in some people. A study from a team at the Weizmann Institute of Science, Israel, found that in people who had taken a mix of antibiotics affecting their gut environment, probiotics inhibited the recovery of the natural microbes by competing with them. This suggested that probiotics are not an ideal substitute for the diverse natural microbial community and after taking antibiotics, for example, it might be better to let the gut recover naturally.

Another consideration is that for those who are well and already have a healthy lifestyle, we don't know whether we gain any

16 Somewhere between 20 per cent and 40 per cent for selected strains.

additional health benefit from consuming probiotics, especially when our bodies already have a good internal regulatory system. After all, what we might consume is just a tiny fraction of the numbers produced in our guts. Just as there are different types of probiotics, there are also different types of prebiotics that are selectively used by different gut microbes. If you don't know which microbes you have that need enriching, you won't know what prebiotic to take that might help. Despite the need for a lot more research to fill the gaps in our knowledge, there is evidence that people with certain conditions have benefited from their use. An analysis of 313 trials of probiotics, involving nearly forty-seven thousand people, found beneficial effects of probiotic supplementation in preventing diarrhoea and respiratory tract infections, and improving measures of inflammation, in people with pre-existing illnesses, for example. However, the quality and comparability of these studies is limited, and the effects were only seen in some participants who already had an underlying health condition. Overall, it seems that there are scientific findings to support the use of probiotics in some people who already have health issues but there's not yet a good case for proactively taking them to prevent ill health. Data on the effects of prebiotics are sparse, and ensuring you consume lots of diverse fibre sources from fruit and vegetables is probably more effective. Dietary fibre appears to strongly influence our gut microbiome and is better related to our health than probiotics or prebiotics.

As usual, while it might not be popular advice, the key is to eat a well-balanced and varied diet. In fact, given the variation in the gut microbes between individuals and the importance of food on the gut's health, research is taking place to investigate whether what we all really need is our own personalised diet tailored to our gut. This could be the future we need to aim for rather than more and diverse wellness drinks that are likely to contain only a fraction of the probiotics found in the pharmaceutical-grade supplements that have shown promise in clinical trials.

A very pretty ingredient that has been lighting up the wellness drinks world is blue-green algae. (It's time to sound the 'superfood claim' claxon). While blue-green algae describes a diverse group of plant-like organisms that grow in salt water, it is particularly *spirulina* that is added to drinks for its supposed health benefits. Manufacturers and health gurus have been using spirulina in beverages, giving them an attractive blue-green colour, as a source of protein, iron and vitamins. It is said to help with conditions such as diabetes, anxiety, depression and premenstrual syndrome, as well as boost the immune system and digestion, improve memory, fight the effects of ageing and increase energy (i.e. the usual things that wellness drinks claim to help with). Interestingly, spirulina and other microalgae are part of a rapidly expanding market into exploiting new and sustainable food crops, for animals and humans, at a time of population expansion and limited arable land and fresh water sources. Microalgae are considered a promising source of nutrition found to contain beneficial properties for health, but how digestible and bioavailable the nutrients actually are in humans is as yet unconfirmed. Although microalgae has promise as a nutrient in the future, the amount you might get in a wellness drink is pretty small and wouldn't add significantly to your intake of protein or vitamins. Unsurprisingly, as blue-green algae drinks are a fairly new addition to the wellness market, there haven't been any studies investigating whether they do in fact have any significant health benefits whatsoever. The feeling by some nutritionists is that although microalgae do contain a lot of nutrients, they don't really provide enough of them in each serving you're likely to encounter, plus they are expensive in how they're currently offered, in comparison with other sources of protein, iron and vitamins.

A number of drinks now offer the added extra of activated charcoal. You may remember that I mentioned activated carbon in the section on water filters. Activated charcoal is the same thing – a special type of highly microscopically porous charcoal used to soak up unwanted particles, not to be confused with burned toast or barbeque briquettes.[17] In medical settings, it has been used for over 180 years as an important means of treating people who have ingested a poison or overdose of drugs. It binds to many drugs or poisons to prevent them being absorbed by the body. In the wellness drinks world, it is peddled as everything from a hangover cure to a beauty aid to a detoxifying agent. The theory goes that if it works to take away nasty drugs from your gut then it must also be able to mop up other nasty toxins in your digestive system. These 'toxins' tend to refer to unwanted things in our diet, rather than things that are actually toxic.[18] The way devotees promote the drink is something along the lines of: *activated charcoal binds to toxins in food, like pesticides and other chemicals, and it is so good at it that medical professionals use it routinely to treat people who've been poisoned.* It all sounds highly plausible, and the mention of its use by medical professionals affords credibility, but there's absolutely no scientific proof that drinks with added activated charcoal work at all. The dose of activated charcoal used for poisonings can be around 100g, followed by further doses every few hours. The amount included in a typical drink is around half a gram. Is this really going to do much? It's pretty unlikely. Manufacturers won't be very motivated to include much more as the activated charcoal may impair the taste and mouthfeel of drinks – something that is more likely to drive down sales than worries over whether it is effective or not. Even if it were actively drawing away molecules, ask yourself how this activated charcoal would know the difference between the nutrients you want to absorb and those that

17 As advised by the National Capital Poison Center in the US (National Capital Poison Center, 'Activated charcoal: An effective treatment for poisonings.' https://www.poison.org/articles/2015-mar/activated-charcoal). Do people really confuse them?!
18 The dictionary definition of toxic is 'poisonous', which is rather over-exaggerating the harms of most of what we eat.

you want to get rid of. The answer is of course that it doesn't know. Activated charcoal is fairly indiscriminate so if a molecule has the right structure to stick to it then that'll happen whether it is a toxin, a medicine or a nutrient. At the low levels typically included in wellness drinks there's unlikely to be much harm from them, although you should probably avoid these drinks if you take medication as the charcoal might render it less effective if it prevents the active ingredients being absorbed. And by the way, for those of you hoping it's a great cure for a hangover you can give up on that hope right now. Activated charcoal is poor at binding with alcohol so in this regard would be pretty useless anyway.

Activated charcoal drinks are not the only ones that lay claim to a detox effect. Plenty of wellness drinks purport to do such a thing, but why are we so obsessed with detoxing? These days we are increasingly aware of the need to maintain a healthy lifestyle in order to stave off weight gain and serious illness and support good physical and mental health. For most of us, who are otherwise well, this means eating a well-balanced diet and getting plenty of exercise, but for many this is either not going far enough or just feels like too much effort. These people may be drawn to products claiming to detoxify the body of harmful chemicals we're exposed to every day or the toxins stemming from lifestyle excesses. But do we really need to 'detox'? Firstly, we don't need to be afraid of terms like 'chemicals', because everything in the world is made up of chemicals, or 'toxins', because in this context they are effectively meaningless. They are just used to instil a sense of fear that there's a problem that we need to fix before selling us a nonsensical product to fix the non-problem. Secondly, for anything our bodies need to get rid of, we already have in-built mechanisms to deal with unwanted substances – handy, given we've been constantly exposed to unwanted substances since the dawn of time. Our gastrointestinal system, lymphatic system, kidneys, liver and skin deal with them very efficiently. Simply put, the idea that we need extra help to remove everyday contaminants in a myth. And while we're on the topic, beware of the word 'natural' to infer that something is healthy. There are plenty of natural things that are highly toxic and can kill you; and plenty of others that have no useful physiological effect in humans.

Soft drinks world on a high

A hundred years ago soft drinks giant Coca-Cola was using a recreational drug[19] in its classic drink and it may be about to repeat this tradition; but this time with a difference. In September 2018, it was rumoured that Coca-Cola was in talks with another company, Aurora Cannabis, to develop marijuana-infused beverages. The aim is to introduce a new type of functional drink to the wellness drinks market that helps alleviate pain. Sources stressed that the addition of cannabis extracts is not about helping users get a high from the drink. This is because the ingredient they allegedly intend to explore is cannabidiol (CBD), the component touted for its medicinal benefits but that does not affect your brain in such a way to produce a high (that's down to a different element – tetrahydrocannabinol [THC] is responsible for most of the psychological effects of cannabis). CBD is already used in pain-relieving medication, amongst other uses, and Coca-Cola is not the only one interested in developing drinks infused with it. The Cannabis Drinks Expo in San Francisco in 2019 provided a platform for drinks manufacturers, and others, to look collectively at ways to utilise CBD and take advantage of a new, and potentially burgeoning, market. We'll have to watch this space . . .

19 Cocaine.

A quick fix?

As you can see, although the theories behind many wellness drinks seem convincing, in most cases there just isn't the hard science to back them up. They aren't the quick fix we wish they were. Much of the 'evidence' comes from personal testimonies and cherry-picked scientific details and hypotheses from small studies in unrelated species and situations. Often the claims run like this . . . *Natural compound A converts compound B into compound C, which is known to have antioxidant/immune-boosting/detoxing etc properties. Therefore, we added compound A in our drink to help you feel better/ look beautiful/live forever etc.* What you, as the consumer, don't get told is that:

a) all of this information is from plant or cell studies, not studies in people;

b) although compound A converts compound B into C, we don't know if there is any compound B in the drink or in your body to convert;

c) compound A may not be bioavailable in humans (i.e. may just pass straight through you without passing on any benefits);

d) a lot of compound A may degrade during the processing and storage stages, leaving very little for the consumer to consume;

e) we don't know if you actually get any compound C after you drink this product;

f) although compound C has been linked with beneficial health effects, if you get any from this drink it is likely to play only a tiny role amongst a host of other compounds and is unlikely to make much of an impact in its own;

g) you have to consume a huge amount of compound A to benefit from its effects but only a small amount is in the drink;

h) most of the drink is made up of other ingredients that are of no benefit whatsoever and may even be of some harm (e.g. sugar).

In other words, just because the drink contains supposedly beneficial compounds it doesn't mean that it will have any tangible benefit when you actually consume it. We don't yet have a wealth of robust evidence from studies of the effects of such beverages in large groups and different populations. If you've already found that these drinks make you feel better then that's great, but I wouldn't advise you to buy them in the belief that they are going to solve your health concerns. It may be that in the future some wellness drinks turn out to be proven health aids but until that time save your money (unless, of course, you're happy to pay inflated prices just for the taste).

CHAPTER FIVE

Alcoholic Drinks

....

Hans Island: an uninhabited, barren rock, lying in the Arctic, around half a mile square and containing no known reserves of oil or natural gas or in fact anything particular of worth. It sounds of little consequence and yet Hans Island is at the centre of a long-running dispute between Canada and Denmark. The desolate island lies in the middle of the Nares Strait that separates Canada and Greenland, an autonomous territory of Denmark. Under international law, countries have the right to claim territory that lies within 12 miles of their shore. As the Nares Straight isn't particularly wide, Hans Island technically falls within both the Canadian and Danish waters. The dispute between the two countries reaches back decades to the 1930s. In 1933, the Permanent Court of International Justice of the League of Nations declared that Hans Island was Danish territory, but these authorities were soon abolished and superseded by the United Nations and International Court of Justice. This means the original judgement carries no weight. The whole issue fell off the radar as more pressing concerns arose, namely the Second World War and the Cold War, but re-emerged in the 1970s. During this time, Canada and Denmark agreed on the demarcation of their maritime borders through the Nares Strait but couldn't agree on what to do about Hans Island, leaving the matter unresolved. The island experienced Danish and Canadian visitors in ensuing years. The turning point came in the 1980s when, after hearing that Canadian researchers had visited the island, the Danish Minister for Greenlandic Affairs flew to the island and reportedly planted a Danish flag and left a bottle of the Scandinavian spirit, aquavit. This marked the start of a polite tit for tat in which the two countries would periodically switch the flags, leave a note welcoming people to either the 'Danish Island' or 'Canada', take the bottle of drink and replace it with their own. It's said that the Canadians would leave a bottle of Canadian Club whisky.

The Danes and Canadians clearly identify with specific alcoholic drinks in representing who they are. In many parts of the world, alcohol is ingrained within national identity. If I named a country and asked you to think of a soft drink from there you might struggle to be specific, but if I asked you to think of an alcoholic drink I bet the task would be a lot easier. When we think of places around the world, we often associate an alcoholic drink with them: England and (warm) beer, Scotland and whisky, Ireland and Guinness, France and wine, Greece and ouzo, Spain and sangria, Austria and schnapps, Russia and vodka, Japan and sake, Mexico and tequila, Cuba and mojito, Jamaica and rum, and the list goes on. I don't think there is a good answer as to why this is, but it's clear that alcoholic beverages are deeply rooted within many cultures, so much so that they become considered something of a national drink and often renowned around the world.

Alcohol: in essence

In 2016, around 2.3 billion people in the world were alcohol drinkers. Of course, the number of people who consume alcohol varies hugely by country. More than half the population in the Americas, Europe and the Western Pacific region consumes alcohol but elsewhere the proportion is much lower. The highest levels of consumption per person are in European countries. However, while alcohol consumption in many parts of the world has remained relatively stable over the last ten years or so, it has been decreasing in Europe, from 12.3 litres of pure alcohol per person in 2005 to 9.8 litres in 2016. Conversely, over the same time period, consumption has increased in the Western Pacific and South East Asia regions.

Alcoholic drinks are simply defined as beverages that contain more than a minimal amount of ethanol/ethyl alcohol. Beer, wine, spirits and other alcoholic drinks all begin with fermentation, which is the natural result of yeast digesting sugars. The yeast eats the sugar and multiplies, producing carbon dioxide and ethanol. Ethanol is the form of alcohol that we drink. The amount of alcohol in drinks is shown on the label and stated as the percentage of Alcohol by Volume (ABV). This is the amount of alcohol in the drink as a whole so, for example, if a drink says 4 per cent ABV that means it contains 4 per cent pure alcohol.

The concept of counting alcohol units was introduced in the UK in 1987. The idea was to provide a way for people to calculate how much alcohol they were actually consuming and keep track of their drinking. One unit is the equivalent of 10ml, or 8g, of ethanol and is the amount that the average adult can metabolise in one hour. In other words, after an hour there should be little or no alcohol left in a person's bloodstream, although this will vary from person to person. The number of units in a drink is based on both its size and its strength (i.e. the concentration of alcohol it contains). For example, a small glass (125ml) of wine at 12 per cent ABV is 1.5 units, a can (440ml) of lager at 5.5 per cent ABV is 2 units, as is a pint of lower strength beer (3.6 per cent ABV), and a single small shot (25ml) of vodka at 40 per cent ABV is just 1 unit. Other countries measure drinking levels in different ways.

Another term once commonly found on bottles of alcohol is 'proof' (usually on spirits) but what does this mean? The term originates from the sixteenth/seventeenth centuries in England when it was introduced as a way of taxing drinks according to their alcohol content. To test, or 'prove', the alcohol content, the tester would soak a pellet of gunpowder with the drink and then try to ignite it. The idea was that if it was still possible to ignite the wet gunpowder then the alcohol level was 'above proof' and taxed at a higher rate. Proof was based on an arbitrary standard of the time, called 100 proof, that was the typical amount of alcohol in distilled liquors. The alcohol content of other drinks was then rated against this, i.e. how much smaller or greater their alcohol content. There were various problems with this method so the measurement eventually became standardised in the nineteenth century based on specific gravity, i.e. the ratio of the density of the beverage to that of an equal volume of distilled water. A 100 proof spirit was deemed to be one with an alcohol content 12/13th the weight of an equal volume of water. This corresponds to about 57.1 per cent ABV. Confusing, huh?! While ABV is a standard international measurement of ethanol content, proof is a measurement that inconveniently varies from country to country, meaning that if you bought a bottle of alcohol in another country or even an imported brand, you could be easily baffled about what exactly you were getting. Beyond the UK, other countries sensibly decided to make things simpler. In the US, for example, alcohol proof is calculated as double the percentage of alcohol in the drink (i.e. double the ABV): 100 proof is the equivalent of 50 per cent ABV, as 50 per cent alcohol is typical of strong spirits. In France, they took 100 per cent ABV as 100 proof and 100 per cent water as 0 (zero) proof to provide a convenient scale. As we now have the standard ABV measurement it rather makes the use of 'proof' on labels fairly pointless and, in fact, it has largely been done away with over the years.

Ethanol tastes different to different people. Some find it very bitter and unpleasant but others may perceive some sweetness with less bitterness. This may explain why some people find drinking alcohol a more pleasant experience than others and why they may be more inclined to drink it. A study published in 2014 identified that these taste differences were linked to different versions of the TAS2R38 gene. Variations in this gene have previously been linked with alcohol intake, with those with the more sensitive form of the gene drinking significantly less. It is also associated with supertasters, who are more sensitive to bitter tastes (see Chapter 4, Cold drinks for more on supertasters). For those who enjoy the broader flavours of alcoholic drinks but are not so keen on the alcohol itself there are now many low-alcohol alternatives on the market for them to savour.

How is alcohol removed from beer and wine?
There is an increasing variety of low- and no-alcohol versions of beer and wine, in response to rising demand for them. Alcohol-free beer contains very little alcohol. In the UK, for instance, it must contain no more than 0.05 per cent ABV. You may think it should have no alcohol at all but this can be tricky to achieve and, to put this into perspective, many other non-alcoholic drinks and foodstuffs naturally contain a similar amount or more alcohol. For example, orange juice can also contain around 0.05 per cent alcohol. There is also de-alcoholised beer (no more than 0.5 per cent ABV) and low-alcohol beer (no more than 1.2 per cent ABV). Low-alcohol drinks refer to those that have an ABV between 0.5 and 1.2 per cent, whereas reduced alcohol drinks are those that have an alcohol content lower than the average strength of that type of beverage. So, for example, a wine that has an ABV of 6 per cent is technically a reduced alcohol drink as most wines are in the region of 11–15 per cent ABV.

Low-alcohol beers start life as regular beer. They go through the usual brewing stages but have their alcohol reduced or removed at the end of the process. This can be done in a number of ways. The beer can be heated to evaporate off the alcohol, as alcohol has a lower boiling point than water, but this can affect the flavour. For this reason some brewers undertake vacuum distilling, which lowers the boiling point of alcohol and subsequently preserves some of the potentially volatile flavour chemicals that might otherwise be lost during heating, helping to retain a flavour closer to the original beer. Another method used to remove alcohol is reverse osmosis. The beer is passed through a very fine filter that only allows alcohol and water, plus a few volatile acids, to get through. The alcohol is distilled out of the alcohol-water liquid then the remaining water and acids are mixed back in with the sugars and flavour compounds that were left on the other side of the filter. As heat isn't used in reverse osmosis, the flavours are less affected by the process. Regular beer becomes carbonated as it ferments within the bottle but this doesn't occur in non-alcoholic beers. Without intervention they would remain flat, so most producers inject carbon dioxide into low-alcohol beers during the bottling/kegging/canning process. Halted fermentation is another process sometimes used to produce non-alcoholic beers. This is where fermentation during the regular beer making process is interrupted before it can really get started thereby limiting the creation of ethanol.

Low-alcohol wine is made from regular wine that has had some alcohol removed. The process for removing alcohol from wine isn't that dissimilar to beer. It may be removed by vacuum distilling or by reverse osmosis. Centrifugal force may also be used to effectively spin the alcohol away from the rest of the wine. This needs to be repeated many times to break down the alcohol molecules.

The rest of this section will now focus on the alcoholic versions of traditionally alcoholic drinks. I'll describe the basics of how various drinks are made but, as you can imagine, there is a lot more to the processes than I have the space to describe. Producing good quality alcoholic beverages involves a precise, and rather magical, world that takes years to truly understand and perfect.

Beer

I went to school in the home town of *Greene King* and vividly remember the distinctive smells from the brewery wafting through the air. As a child, I couldn't stand the smell but I suspect that I would now have a different appreciation for it. There are probably many others who can't get enough of the aroma and what it represents because, after all, people around the world love beer and have been making and drinking it for thousands of years. The earliest evidence of beer-making discovered so far dates to around thirteen thousand years ago from a site in Israel. While there was likely little resemblance between that beer and today's beers, the same basic principles for its production were clearly established long ago. Beer is a drink made with malted grains (particularly barley but also potentially wheat, rye, corn or rice), hops and water that is fermented by yeast. Alcohol levels in today's beers range considerably from about 3 per cent ABV to 10 per cent ABV, although most are between 3 per cent and 6 per cent.

How is beer made?
Brewers use a range of ways to produce their beer but there are certain general methods common to them all. After the grain is malted (see Chapter 3, Hot drinks to find out more), it is heated with water in a process known as mashing. This breaks down the starches in the malt into simple sugars. The sweet liquid drained off is called the wort. Hops, and any additional spices desired, are added to the wort and the whole thing is then boiled. Hops are the flowers, or seed cones, of a climbing vine. They impart bitterness to the sweet wort and help to provide balance and depth of flavour to the beer. They also act as a natural preservative, helping extend the life of the beer. The wort is then cooled and filtered, and yeast is added to it. The brewing is complete at this stage and the fermentation stage takes over. The beer is stored for a period of time while fermentation takes place after which it is either bottled up or transferred to a conditioning cask to age. At this stage the beer is flat. Some beer will have carbon dioxide added to create the fizz, whereas others are left to age where additional fermentation takes place creating natural carbonation.

Beer types
Beers are categorised as either ales or lagers, depending on their fermentation process. Ales are the oldest style of beer, tracing back thousands of years. During fermentation, they are stored for a couple of weeks at room temperature. Lagers are fermented at cooler temperatures and for many more weeks than ales. Different types of yeast are

used that ferment at different temperatures. While there are a number of yeast categories, the main two are top fermentation yeasts and bottom fermentation yeasts. Top fermentation yeasts are used in making ales, porters, stouts and wheat beers, among others. The yeast settles at the top of the beer. Bottom fermentation yeasts are used to make lagers and settle at the bottom of the beer. There are also a few beers that are produced by spontaneous fermentation, where no yeast is added but the beer is exposed to the open air to allow wild yeast and bacteria to infect it. These are lambic beers, a speciality in some parts of Belgium, said to have a sour, and acquired, taste. Ales are darker in colour, heavier and more bitter than other beers, whereas lagers are typically light in colour and are more effervescent.

There are many types of ale. Indian Pale Ale (IPA) is a strong and bitter beer. It was allegedly born out the need to preserve the beer during long journeys overseas. The extreme temperatures and prolonged storage supposedly wreaked havoc with beer that was being sent to far-flung parts of the British Empire. To combat this, the beer was loaded with extra hops to confer their preservative magic, and at the same time they imparted additional bitterness and depth. Many IPAs exist, from around the world, with different strengths and herbal and citrus elements. Pale ales are also hoppy beers but milder and pale and golden in colour. Bitters grew out of pale ales. They are generally darker in colour than pale ales as slightly darker malts are used. Best bitter is a stronger version of this style, although this varies. Mild is a traditional style of beer, less hoppy than bitter, with a dark brown colour and hints of chocolate and nuts. Stouts and porters are much darker beers with a strong flavour, and often have a higher alcohol content. Guinness is arguably the world's most famous stout. There's a lot of overlap between the two categories but in general it tends to be the case that porters have a less roasted and so milder flavour than stout. There are a number of other differences stated in beer circles, but these are often contested and it seems no one can agree on definitive, distinctive, classifications of the two beer types. Wheat beers, as their name suggests, are made with a large proportion of wheat, along with the malted barley. They are made using specialist yeast. They are cloudy in appearance as a result of the proteins in the wheat (of which there are more than in barley) and yeast that are not filtered out of the finished beer. While they are a diverse group, wheat beers are typically effervescent, fairly light in flavour, low in hops and can taste relatively sweet and fruity.

Moving on to lagers, these range in colour and bitterness but most are a light golden colour, with a medium to high hop flavour, and high carbonation. Globally, lagers are the best-selling beers. The most well-known lager type is probably pilsner. Pilsners, originating from the Czech Republic, are crisp, highly carbonated, golden lagers that are refreshing and easy to drink. Other lagers include pale types, like bock beers, which are more hoppy than other lagers, and helles lagers that are similar to pilsners but less sweet, and dark types, such as dunkels and schwarzbiers, that are very malty and may have coffee and chocolate flavours.

What I've described above is a very broad categorisation of beers. In reality, there are dozens of beer types and styles, with different parts of the world and different breweries interpreting them in their own way. For example, you would expect beer crafted in Belgium to taste quite different from that from the UK, and both would be different again from beer made in the US or China.

Beware the beer goggles

If you've not heard the term before, you are wearing your 'beer goggles' when the consumption of alcohol alters your perception of how attractive someone is. Generally, this phrase is used when we find someone to be more handsome or beautiful after we've drunk alcohol than we would actually perceive them to be when sober. We might find someone attractive one evening then the next morning re-evaluate our judgement, possibly

---

have a momentary shudder in horror, and blame it all on the beer goggles. The term has been around for several decades but while the phenomenon doesn't happen for everyone, researchers have actually determined a scientific explanation for it. Scientists at Roehampton University and the University of Stirling recruited volunteers, who were either sober or intoxicated, to observe images of faces and judge how symmetrical they were and also state which were most attractive. Facial symmetry, amongst other factors, is something that our brains perceive as being important in how attractive someone is. The theory behind the study was that alcohol might impair our ability to detect this symmetry and therefore render us less able to appraise attractiveness critically. This is possibly due to reduced visual acuity. The findings of the research bore this out with sober participants being better able to determine whether faces were symmetrical or not, as well as expressing a greater preference for symmetrical faces. A second study by the same team found further evidence to support these findings. These were only small studies but they offer some insight into how the beer goggles phenomenon may occur.

An interesting twist to this is some research that found that the beer goggles effect may even work on oneself. A winner of an Ig Nobel Prize[1] in 2013, a study from researchers in France and the US discovered that people think they are themselves more attractive than they actually are when they've consumed alcohol, even though sober observers don't agree. The more alcohol the study participants drank, the more attractive they thought they were.

[1] The Ig Nobel Prizes recognise research achievements that 'make people laugh then think'. They are intended to celebrate the unusual and the imaginative, and stimulate people's interest in science, medicine and technology. (Improbable Research, 'About the Ig Nobel Prizes.' https://www.improbable.com/ig/)

So, for some people, beer goggles can make others seem more attractive and themselves feel more attractive. It all sounds very uplifting . . . until they sober up, of course!

Cider and perry

On a balmy summer evening, a crisp cool cider can really hit the spot and offers a refreshing alternative to other alcoholic drinks. Cider is made from fermented apple juice. (In the US, cider just refers to unfiltered, unsweetened apple juice, not the alcoholic drink, and can lead to confusion among people from other countries, myself included. Hard cider is the American term for the alcoholic version.) Perry is a similar drink made from fermented pears. Cider contains a similar alcohol content to beer, ranging between 4 per cent and 8 per cent ABV.

How is cider made?

There are hundreds of cider apple varieties, each with its own character, that will affect the final flavour. Cider apples are generally categorised in four groups: bittersweets (with low acidity and high tannin), sweets (low acidity, low tannin), sharps (high acidity, low tannin) and bittersharps (high acidity, high tannin). By selecting and combining particular apples, cider producers can create blends suited to their requirements. Just as we think of wines being good in particular years due to the weather and conditions the grapes experienced, environmental elements can also affect the quality and taste of apples, which in turn will affect the resulting cider.

After the apples are washed and pulped, they are pressed to release their juice and this is then fermented. It often undergoes two types of fermentation (although the second stage isn't always carried out). During the first fermentation, yeasts, that are either added or naturally present on the apple skins, convert sugars to ethanol and other alcohols. Secondly, something called malolactic fermentation takes place, during the maturation phase. This is where bacteria convert malic acid to lactic acid and carbon dioxide. Malic acid adds a sourness to products, so if there is high acidity this fermentation stage can help to better balance the flavour, as lactic acid is a smoother, more mellow, buttery flavour. Fermentation may take place in bottles or in tanks. If it occurs in bottles, you'll notice a sediment at the bottom, which is the remains of the fermentation yeast. Tank fermented cider will be clear as it is syphoned off from the tank, leaving the yeast behind. Once the cider has fermented, additional flavours, such as sugar or other sweeteners, fruit juices or spices, may be added. The cider may be filtered to remove unwanted debris.

Cider types

There are different styles of cider, including still or sparkling, dry or sweet. Ciders often develop natural carbonation during the fermentation process. Following filtration, the carbon dioxide may be lost from the liquid. The result may be served as a still cider or be specifically re-carbonated to provide the fizz that many people enjoy. Not all ciders are filtered, however, and these may be enjoyed with their natural bubbles and are considered finer ciders. Tannins in the apples impart both colour and bitterness to the cider, so, for example, those based on apples with a higher tannin content will be darker and drier than other ciders. This is the essential starting point for the flavour of cider. After fermentation, the result is an unsweet 'dry' beverage, where the mouthfeel is as though moisture has been drawn from your tongue, as the sugar has been converted into alcohol by the yeast. This can then be sweetened to achieve the desired end flavour and is why ciders range from the very dry to the very sweet.

Wine

Vineyards are wonderful places, often in sunny climates and beautiful locations, and visiting them is quite a treat. I've been lucky enough to tour vineyards and wineries in New Zealand, California, France and England, and a chance encounter with a wine buyer in Barcelona unexpectedly expanded my taste for Spanish white wine. My favourite memories from some of these include: exploring a fascinating biodynamic vineyard in Sonoma Valley where I learnt more about bats than wine; a diminutive elderly wine seller accidentally kicking over an enormous bottle of wine in Châteauneuf-du-Pape and not even noticing; the eccentric tour driver in New Zealand who insisted on taking a photo of me and my partner in front of the iconic Cloudy Bay mountain, only to capture the mountain and a slice of my partner, cutting me out entirely. And, finally, I can't not mention the brilliant, and unforgettable, couple who reminded us of Neil and Christine Hamilton,[2] where the husband believed himself to be something of a Malbec expert only to be contradicted at every turn by the wineries, and the wife who said she didn't see the point of wine tours because 'everyone just drinks champagne these days, don't they?'! Good times.

Wine is made from fermented grapes and the alcohol content ranges from about 8 per cent ABV to 14 per cent ABV. Fortified wines contain much more alcohol than regular wine, at around 16 per cent to 22 per cent ABV.

How is wine made?

Vineyards around the world manage to produce a huge diversity of wines because of the many variables that may be manipulated or naturally occur. Different types of grapes, different climates, weather and other environmental conditions, individual growing seasons, care of the vines and the timing of the harvest all affect the wines that are ultimately produced. Slight changes in any of these

2 The former UK politician and his wife who became eccentric TV celebrities.

parameters may affect the level of alcohol, sugar content, acidity and flavours in the wines, for example. There's far too much to say about grapes and how they are grown to do this topic justice here, but it's well worth seeking out a comprehensive book about wine if you would like to know more about this. Let's instead start at the point where the grapes are harvested.

Grapes are picked from the vines and sorted for quality. They are destemmed and then crushed. Grape stems are high in tannins and leaving them in during fermentation would result in an unpleasant bitterness, so they are removed. Wines do contain tannins still but these are primarily from the grape skins. (Having said this, stems are sometimes left on or added back in if additional tannins and flavours are desired.) In days gone by, men and women at the vineyard would take off their shoes, roll up their trousers or skirts and get stomping, squelching the soft grapes under their feet, the skins and juice oozing between their toes. A messy business. These days, for the most part, mechanical pressers are used instead to release the grape juice and get the wine processing going. As the grape skins are broken, the juices come into contact with them, absorbing their tannins, flavours and colours, as well as with the yeast on the skins' surface and from the air. This kicks off the fermentation. It is at this stage that a major difference between wines is determined – whether it is to be red or white wine. For white wine, the crushing and pressing of the grapes is a rapid process that minimises contact between the juice and skins to avoid unwanted colour and tannins from seeping into the wine. So after the grapes are crushed, the juice is then quickly pressed out and away from the skins and other parts of the grape. Conversely, the juice for red wine is left to mingle with the skins to absorb the desired effects. Rosé wines are exposed to the skins for a short period to pick up some of the colour but not the same depth of flavour as red wine. This all refers to using red grapes, which may produce red or white wine. White grapes only produce white wine. This juice-skin mixture is called the must, and either flows, or is pumped, into fermentation vats.

The must is fermented into wine and fermentation may begin as a soon as six hours after the grapes are pressed. Wild yeasts from the skins and air will kick off this process, although many producers will add a commercial yeast to ensure a controlled and consistent process. Fermentation continues until all of the sugar is used up (although with sweet wines, fermentation will be stopped early before all the sugar is converted). Fermentation takes anywhere from six days to a month or more. White wines are fermented at lower temperatures than red wines, which helps achieve the fresher and more aromatic flavours. After fermentation, the solids in the wine (i.e. the skins) are removed and then pressed to extract all the liquid. This results in 'pressed wine' that is matured separately and may or may not be blended in with the other wine at a later stage to add flavour and colour. This pressing stage takes place prior to fermentation, straight after the crushing stage, for white grapes. Most red wines and some rich white wines (e.g. Chardonnay) also undergo malolactic fermentation (see *Cider* for more on this), giving them a richer, velvety texture and softer, less acidic flavour.

Fermented wine is moved to another container (e.g. bottle, stainless steel tank or oak barrel), leaving behind the sediment created by the dead yeast and other grape debris. This process is called racking. The dead yeast is referred to as lees and while it is typically removed, some wine makers choose to leave in some fine particles (particularly in white and sparkling wines) for different lengths of time to add to the flavour and mouthfeel of the wine. The lees impart a buttery, creamy, toasty richness to the wine, adding to its complexity. So, if you see 'sur lie' on the label, it means the wine has been aged 'on the lees'. Racking also exposes the wine to some aeration which aids the wine's development.

Wine can then be bottled up or left to mature and then bottled. Many wines are aged in wooden (typically oak) barrels as the wood affects the flavour, imparting smooth hints of toast, vanilla or coffee. A similar effect can be achieved by adding oak chips into wine held in steel tanks. Some wine producers are now using cement or clay vats to age their wine in, hoping to add mineral characteristics. Steel tanks are typically used for zesty white wines.

Wine also undergoes clarification (either before maturation or prior to bottling) through fining or filtration. These processes clarify the wine and remove any solids and unwanted debris. Filtration involves filtering the wine through increasingly fine filters until it looks bright and clear. However, some wine makers feel this strips the wine of its unique properties. Fining is the reason why many wines have allergy warnings on their labels. It is the clarification process where substances are added to stick to unwanted particles in the wine and force them to the bottom of the tank or barrel. They also help reduce astringency and reduce colour. Substances used by wine makers to do this job include: egg white, gelatine, isinglass (a preparation of the collagen protein from fish bladders), casein (from milk), chitosan (made from chitin from shellfish), carbon, bentonite (a type of fine clay made from aluminium silicate), kieselsol (made from silica gel or silicon dioxide), and polyvinylpolypyrrolidone (a synthetic polymer). If you've heard about vegan wine and wondered what on earth that's all about given wine is made from plants, it's where the wine has been produced without fining substances that originate from animals. The clarified wine is then transferred into another container for either storage or bottling.

Something else you might be warned about on a wine label is the presence of sulphites. It might even be something you're concerned about (more on the health effects of sulphites shortly). Sulphur dioxide, a type of sulphite, is used in wine making to help preserve it by inhibiting the growth of yeast and bacteria that could cause the wine to spoil. Sulphur dioxide occurs as a by-product of fermentation but most producers also inject additional sulphur dioxide. It is the added sulphur dioxide that is referred to on a label stating the wine 'contains sulphites'. There are some wines, such as organic wines, that have not had sulphites added to them but they will still contain some natural sulphur dioxide. Sulphur dioxide is also often used to clean barrels and a small amount may end up in the wine stored inside them.

Sparkling wines are produced by a number of methods, each resulting in a different amount of carbonation and style of drink. They all use a base wine (usually white). The first is the Champagne method (also called the traditional method outside of the Champagne region). It is typically the most complex and most expensive process. It requires secondary fermentation to occur inside the bottle when something called *liqueur de tirage* is added to the wine. *Liqueur de tirage* is a mixture of wine, in which sugar has been dissolved, and yeast that when added to the base wine kicks off secondary fermentation, leading to the formation of carbon dioxide bubbles released by the yeast. The bottle is temporarily capped while the secondary fermentation works its magic. The wine ages (on the lees) during this time and, in the case of Champagne, may take anywhere from fifteen months to several years to mature. This results in a richer flavour and mouthfeel than some other sparkling wines. When ready, the bottle is inverted so the lees fall to the cap and the bottle neck is frozen. The cap is then removed and the pressure inside the bottle forces out the plug of sediment. As a small amount of wine is lost, the bottle is topped up with reserve wine along with sugar (called dosage), then the final cork added. The amount of residual sugar remaining in the wine plus the sugar that is added at this stage determine the type of Champagne, from 'Brut Nature' (no added sugar and little residual sugars) to 'Doux' (more than 50g per litre of residual sugars). Champagne, Cava and Crémant are all examples of sparkling wines made via this method.

The tank method (also known as the Charmat method) is used in the creation of Prosecco and Lambrusco and is a less expensive process than the Champagne method. Rather than fermenting each bottle separately, this method carries out the secondary fermentation in a large tank *en masse*. When ready, the wine is filtered and bottled without extended contact with lees. This results in lighter, fresher wines with larger bubbles than the Champagne method. The transfer method is largely the same as the Champagne method but instead of removing the lees from the individual bottles, the wines are emptied into a pressurised tank, the sediment filtered off and the wine then put into new bottles. There are also methods called the ancestral method (that uses very cold temperatures and pauses fermentation midway through the process) and the continuous method (aka Russian method, that continuously adds the *liqueur de tirage* to the base wine that is pumped through a series of tanks). Finally, an inexpensive way of producing sparkling wine is simple carbonation; i.e. taking base wine in a tank and injecting it with carbon dioxide. Sparkling wine made in this way is considered inferior, not least because the coarse bubbles dissipate quickly, and often cheap bulk wine is used to produce it.

Wine types

As you'll know, there is a lot more to categorising wine than just red, white, rosé and sparkling. I've talked about some of the ways in which red, white and rosé wines get their colours and flavours. There are hundreds of different grape varieties, leading to a wide variety of wines, but there are a number of wine types that are particularly popular that I'll mention. Wine may be named after the principal grape variety used to make it or after the region it was made in. This can be confusing as it means that the same type of wine might be labelled as different things depending on where the wine came from. In general, wines hailing from the New World (e.g. Australia, New Zealand, South Africa, the Americas) are named after the grape variety, whereas many Old World wines (i.e. Europe) refer to the region. For example, a red Burgundy from France is Pinot Noir and a white

Burgundy is Chardonnay. Some wines are actually a blend of more than one grape so the label will usually refer to the grape variety that makes up the greatest proportion of the wine.

On the whole, red wines have a more robust, fuller flavour than white wines and are usually drunk around room temperature. They range from full bodied and somewhat savoury, such as Cabernet Sauvignon and Malbec, to light bodied and fruity, such as Pinot Noir and Beaujolais. Types like Syrah, Grenache, Merlot, Montepulciano and Zinfandel come somewhere in between. White wines range from dry and rich, such as Chardonnay and Viognier, to light and sweet, like Riesling, with types such as Sémillon, Muscadet, Sauvignon Blanc and Pinot Grigio (Pinot Gris) falling in between. Rosé wines range from dry to sweet and can be created from a range of grape varieties. Classic dry types are found in the South of France, amongst other places, and are often made from Grenache, Syrah and Mourvèdre grapes, whereas sweeter types can be found in the New World, such as those made from White Zinfandel in places like California.

I've already covered something of some sparkling wine types but I must also mention dessert and fortified wines. Dessert wines are very sweet wines that are typically served with a dessert to complement the flavours. They are commonly white wines, although there are some red examples. They are usually produced from late-picked, very ripe grapes that have been deliberately left on the vine so the sugars concentrate. Some wine makers capitalise on a fungal infection of the grapes (*Botrytis cinerea*) that dessicates the grapes and concentrates the sugars and acids. There is also an unusual type of dessert wine called ice wine, which is made by using grapes that have been left to shrivel on the vine until late in the year and then harvested after the first hard freeze. The water in the grapes is frozen and retained in the grapes and the sugars, acids and aromas become concentrated as the grapes are pressed when still frozen. (Although not usually one for a sweet wine, I've been lucky enough to try a

Canadian ice wine and it was amazing.) In parts of Europe, grapes may be laid out on straw mats to dry to concentrate the flavours before pressing.

In addition to the grape ripeness, fermentation can be stopped early to prevent the yeast from converting all the sugar into alcohol. This is done in the case of fortified wines, like port and sherry.[3] To halt fermentation, a spirit (usually grape brandy), is added to the wine. This increases the alcohol content as well as ensuring a significant quantity of sugar remains in the wine. Fortified wines have a longer shelf life after opening than other wines.

Spirits and liqueurs

Spirits encompass a wide range of alcoholic beverages with a high alcohol content. They are produced through a combination of fermentation and distillation; ethanol is distilled from grains, fruit or vegetables that have undergone fermentation. This results in a lower water content and higher concentration of alcohol. Liqueurs are spirits that have been sweetened and usually flavoured. They're sometimes known as cordials. Popular examples include: Advocaat, Amaretto, Bailey's Irish cream, Campari, Cointreau, Grand Marnier, Kahlua, Limoncello and Ouzo. Spirits and liqueurs vary widely in their alcohol content but generally contain over 20 per cent ABV. Vodka, whisky, rum and gin are around 40–60 per cent ABV, and absinthe ranges between 55 per cent and a whopping 90 per cent ABV.

What differentiates spirits from one another is the base ingredients used, and processes employed, to make them. Whisky (or 'whiskey' in Ireland), gin and vodka are all based on starchy carbohydrates; i.e. grains or potatoes. Whisky is usually made from malted barley, although sometimes wheat, rye or corn may be employed. Gin can

---

3 It's worth noting that sherry and port, while both fortified wines, are produced in different ways that I don't have space to cover here.

be produced from any grain then infused with assorted herbs, particularly juniper. Vodka is typically made from grain but is also often made using potatoes. Fruits and natural sugars are the basis of spirits such as brandy, tequila and rum. Brandy is based on fruit, most commonly grapes, tequila stems from agave juice, and rum is made from molasses or sugar cane juice.

Once the base of the spirit has been fermented it moves on to the distillation process. As mentioned earlier, alcohol evaporates at lower temperatures than water and this fundamental principle is why distillation works. Upon heating, ethanol is evaporated off the spirit base (or wash), captured and cooled. The cold temperature condenses the ethanol back into a liquid. Distillation is either carried out in a copper or stainless steel pot still (also called alembic still) or a column still, each of which will affect the flavour and body of the product. Pot stills are big kettles in which the wash is heated. Ethanol evaporates off into a cooling tube that transfers the alcohol into another vessel to condense. The temperature rises during pot distillation resulting in different compositions of the resulting condensate at the beginning, middle and end of the process. Using a column still, on the other hand, allows you to distil the wash multiple times at a constant temperature, leading to a purer, less flavoured, more uniform (i.e. less distinct) product. In this method, column stills are arranged as two stacks, with one column acting as the distiller and the other as a condenser, and contain a series of plates with holes in them. The wash is continuously injected into the column from the top and steam rises up to meet it from a boiling kettle below, taking along the alcohol, out of the wash, as it rises up through the column. Column stills are better for large-scale production as pot stills work on a batch-by-batch basis and must be regularly cleaned out. As column stills are continuously functioning and involve multiple columns at a time, they require no cleaning and allow for efficient, repeated distilling. Spirits like whisky, brandy, rum, tequila and other more richly flavoured spirits are more likely to be produced in a pot still, whereas clear spirits such as gin, vodka and white rum will typically arise from column stills.

During the distillation process, compounds called congeners can also evaporate along with the ethanol. Examples of congeners are tannins, esters, methanol and other alcohols, and these naturally occur as a result of fermentation and distillation processes. These can affect the flavour and mouthfeel of the end product. Some of these might be desirable, whereas others may be less so. The trick in successful distillation is to achieve the right balance of ethanol, desirable congeners and flavour compounds. Precise temperature and timing control is used to separate out unwanted congeners. A good example is the removal of methanol, a toxic alcohol dangerous to health. Luckily, its boiling point is different from that of ethanol so it can be separated relatively easily during the distillation process. Column distillation typically results in fewer congeners than pot distillation.

The end product of distillation is usually colourless and a little coarse in flavour. It needs to be either filtered to purify the product or aged in barrels to bring on colour and depth of flavour. Clear spirits, like gin, vodka, white rum and some brandies and tequilas, are not aged but are filtered to remove impurities and then diluted with water until the desired alcohol level is achieved. The filtering process will affect the overall flavour and mouthfeel of the resulting drink. Drinks like gin are infused with flavours, such as botanical elements, prior to distillation. Whiskies, most rums and brandies, and some tequilas are usually aged in a wooden cask or barrel, often for many years. These containers mellow the spirit and impart flavours like caramel, vanilla, oak, toast or tannins. The type of wood the barrel is made from, whether it is new or used, and what was previously aged in it will all make a difference to the flavour. Some distillers choose to use barrels that previously held sherry or port to reap some of the remaining essences. The longer a spirit is aged, the richer and more complex the flavour. Other elements, such as evaporation of alcohol and oxidation (where flavour compounds react with oxygen, modifying them), also affect the end product.

Don't blame it on the sunshine . . . blame it on the moonshine

Moonshine has a long history. In centuries gone by, it was an illegal alcohol aimed at avoiding the taxman, and indeed it is still illegal in many countries to make moonshine for personal consumption and without a licence. There are a few definitions but, nowadays, moonshine is often referred to as unaged whisky. In countries like the US, moonshine is a drink that is deliberately made as a particular product by various distilleries, and it is sometimes called white whiskey (referring to the fact that it hasn't developed colour through ageing). In the past, the law clamped down on moonshine for tax reasons but also because it was infiltrating the spirits market with poor quality liquor. These days, there are good health reasons why strict rules on the production of moonshine exist. Simply put, moonshine in the wrong hands can be a killer. This is due to the potential for methanol to be contained within it, as a result of poorly controlled distillation. Methanol is a toxic alcohol found in domestic and industrial agents (e.g. perfume and antifreeze). It breaks down into the toxins formaldehyde and formic acid. Consuming methanol can lead to a range of very serious problems, including permanent blindness, Parkinson's-like disease, organ failure, coma and death. In fact, hundreds of deaths from home-made spirits, like moonshine, have been reported in different countries over the years. Without expertise, home distillation is a very risky business, akin to playing Russian roulette.

Cocktails and alcopops

Cocktails are popular alcoholic beverages as they are easy to drink (with all the sweet ingredients that are usually added), invoke positive feelings (images of parties and celebrations rather than someone drowning their sorrows spring to mind) and can impart an air of sophistication on the drinker (here, I'm really thinking about a Whisky Sour or Martini rather than a Sex on the Beach or Chunky Monkey). Cocktails combine spirits with other ingredients, such as fruit juice, soda, cream or flavourings. The spirits I mentioned previously (whisky, gin, vodka, brandy, tequila and rum) form the basis of the majority of cocktails. They have been created and refined over centuries but many of the classic cocktails that we know and love today stem from the nineteenth and twentieth centuries.

Every year, *Drinks International* publishes a list of the world's best-selling cocktails. Despite seasonal trends, many classics remain firmly in the top ten. Here's the list for 2019:[4] any of them take your fancy?[5]

1. *Old Fashioned* – whisky or bourbon, Angostura bitters,[6] sugar syrup, orange
2. *Negroni* – gin, sweet vermouth, Campari, orange peel
3. *Whisky Sour* – bourbon, lemon juice, sugar syrup, egg white
4. *Daiquiri* – white rum, sugar, lime juice

4 Drinks International, 'The world's best-selling classic cocktails 2019.' https://drinksint.com/news/fullstory.php/aid/8115/The_World_92s_Best-Selling_Classic_Cocktails_2019.html?current_page=5

5 The ingredients for these cocktails vary as people put their individual twists on them but this list includes the common ingredients used.

6 While you've probably heard of it, you may not know what is in Angostura bitters. It contains a blend of a wide array of spices, fruits and vegetable extracts, including a bitter root called gentian, and alcohol. It aims to balance acidity and enhance flavour in drinks.

5. *Manhattan* – bourbon, sweet vermouth, Angostura bitters
6. *Dry Martini* – gin, dry vermouth, lemon twist or olive
7. *Espresso Martini* – vodka, coffee liqueur, double espresso, sugar syrup, coffee beans
8. *Margarita* – tequila, triple sec, lime juice
9. *Aperol Spritz* – Aperol, Prosecco, soda
10. *Moscow Mule* – vodka, lime juice, ginger beer

The huge resurgence of gin, and subsequently gin-based cocktails, in the last couple of years may influence this list in the next year or so. In the UK at least, gin has become hugely popular and consumers are seeking expensive brands with interesting botanical blends. Sales have tripled since 2009 and gin distilleries have popped up everywhere.

The popularity of cocktails has come and gone multiple times over the decades, often influenced by culture, such as movies and television. At the top of the list of favourites, the Old Fashioned became popular again in recent years, possibly due to a fondness for it by the lead character, Don Draper, in the TV hit *Mad Men*. The series oozed style and sophistication and the cocktail was lent the same reputation by association. The Vodka Martini has the same feel, thanks to its link with the most famous of spies, James Bond. The Cosmopolitan became the ladies' drink of choice thanks to its prominence in *Sex and the City*. The glamour of early twentieth century Hollywood also prompted a rise in the popularity of particular cocktails as people wanted to emulate their idols. The Bronx, one of the most popular cocktails in the 1930s, featured in *The Thin Man*, the French 75 was popular in 1940s America after featuring in *Casablanca*, the Manhattan was shaken up in a hot water bottle in *Some Like it Hot* and the Gibson appeared in *North by Northwest* and was also favoured in *All About Eve*.

Cocktails in the past were the wellness drinks of today, although the two are now trying to merge; the so-called Superfood Cocktails (*rolls eyes and sighs*). These days, just like the wellness drinks, the world of cocktails is prone to quirky fashions when it comes to ingredients. For example, activated charcoal (that you're probably sick of hearing about in this book) is now being included in some cocktails, both to provide a dramatic black colour as well as to convince the consumer that it is somehow healthy and will soak up the alcohol as well as avoid a hangover. All nonsense, of course. Other ingredients, such as pomegranate juice, kombucha and coconut water, are also being used to offer a sense of healthfulness, although I suspect the effect of the alcohol on your body would counteract any tiny benefit from a splash of green tea or dash of turmeric.

Stirred not shaken?

There are a number of theories regarding the origin of the martini cocktail but most hail back to nineteenth-century America, although an Italian maker of vermouth started using the name 'Martini' back in 1863 which may have something to do with it. Different types of martini exist, including those made with gin, or vodka, or gin and vodka, and they continue to remain among the most popular cocktails. It is such a classic that it even has a day of the year – National Martini Day is 19 June, in case you feel like participating.

Martinis were particularly *de rigueur* in the 1950s and 60s and thrown into prominence as the favourite tipple of James Bond. He famously liked his martini 'shaken, not stirred' but some people question this mixological method. It seems that shaking the cocktail would result in a number of less than desirable qualities. First, that this would melt the ice quicker and dilute the drink; second, that the drink would become cloudy; third, that the alcohol is apparently in some way 'bruised' by the shaking affecting its flavour. It's said that the main purpose of shaking the drink is to increase its contact with the ice and chill it more rapidly. Amusingly, a tongue-in-cheek study of shaking versus stirring martinis, published in the *British Medical Journal* in 1999, found a higher level of antioxidants in the shaken versions, and concluded that 007's excellent health may have been due in part to compliant bartenders! In practice, martinis are usually stirred, not shaken.

Hooch, Bacardi Breezer, Smirnoff Ice

These are just some of the alcopops that changed the drinking habits of the youth. An alcopop is an informal UK term (blending the terms alcohol and fizzy pop) for a ready-made drink that resembles a soft fizzy drink but contains alcohol. These drinks were all the rage from the mid-1990s to the early 2000s. They were visually appealing, highly gluggable and conveniently pre-mixed, ready to drink straight out of the bottle. Alcopops contain a similar amount of alcohol to beer, ranging from 3 per cent to 7 per cent ABV, along with a huge amount of sugar. The alcohol in these drinks may be based on a number of options, including vodka, rum, schnapps, whisky and beer. This is then typically combined with carbonated water, sugar and sweeteners, acidity regulators, flavourings, colourings and preservatives. It's tricky to find out what is in many particular alcopops as alcoholic drinks do not have to have their ingredients and nutritional content listed, like other foodstuffs do.

The popularity of alcopops dwindled as campaigners highlighted their appeal to children (being colourful, eye-catchingly packaged, sweet and not really tasting of alcohol) and their role in the rise of underage drinking. As alcopops became associated with very young drinkers, others turned their back on them and sought more sophisticated alternatives. This impact on consumer habits, along with increased government taxes on alcopops, led to their demise. Although some are still available today, the market for alcopops has reduced considerably since their heyday.

Mixers

Although not alcoholic themselves, mixers are commonly added to alcohol for easier drinking. Mixers are typically soft drinks so I won't cover them in much detail here as I've dealt with that category elsewhere in the book. The most popular are juices, club soda, tonic water, ginger ale, bitter lemon and sweet carbonated drinks like lemonade and cola. Their inclusion in drinks changes the nutritional composition of them and could affect how the compounds in the alcoholic drink interact with the body. A few mixers, notably tonic water and bitter lemon, contain something called quinine. Quinine comes from the bark of the cinchona tree, found in tropical countries, and was used in medicines, particularly for malaria, for centuries but why did it end up in soft drinks? Quinine has a distinctive bitter taste, and the story goes that back in the mid-nineteenth century, officers in the British Army in colonial India would mix their quinine medication with soda water and sugar to make it more palatable. This gave rise to the name 'tonic water' and it was soon married with gin to improve the drink further. The flavour combination, with the bitter quinine re-balancing sweet ingredients, took off ... and the rest is history. Quinine can induce some unpleasant side-effects, like headache, fever, stomach discomfort and tinnitus, but thankfully the amount contained in modern mixers is pretty low and unlikely to cause problems. Bitter lemon became an extension of tonic water, with added lemon juice.

Alcohol and health

I think it's safe to say that drinking too much alcohol is not great for us. Alcohol affects all parts of the body, from our brain and nervous system, and liver and lungs, to our heart, eyes, mouth and skin, and everything in between. Ethanol quickly passes from our mouth to our stomach and into our bloodstream and then our organs. Absorption into the bloodstream occurs within minutes of ethanol reaching our stomach and the level of alcohol in the bloodstream peaks around forty-five to ninety minutes later. Ethanol in the blood is poisonous at high levels so the body needs to break it down and get rid of it quickly. Ethanol is broken down by several pathways, the most common of which involves the enzymes alcohol dehydrogenase (ADH) and aldehyde dehydrogenase (ALDH). These enzymes break the ethanol molecules apart so that they can be more easily processed by the body. First, ADH converts most of the ethanol in the liver to acetaldehyde, a toxic by-product known to cause harm, but this doesn't last long as it is then quickly metabolised by ALDH into another, less toxic compound called acetate. Acetate is eventually broken down into carbon dioxide and water and excreted from the body. Some alcohol metabolism takes place outside of the liver, including the pancreas, the brain and gastrointestinal tract, exposing tissues and cells to potential damage from acetaldehyde.

Your body can only break down a certain amount of alcohol per hour. This amount differs widely between individuals, and some people are more susceptible than others to the effects of alcohol. The individual variations in the way people metabolise alcohol are controlled by a combination of genetic and environmental factors, that mean some are at a greater risk of harms from alcohol consumption. For example, people carry different variants of the ADH and ALDH enzymes that work more or less efficiently, so if you had a fast ADH enzyme or a slow ALDH enzyme, or both, this could lead to a build-up of acetaldehyde and greater exposure to, and subsequent risk of damage from, this toxic metabolite. Other differences between people that may affect their response to alcohol include their liver size and body mass, as well as other genetic factors.

There's no doubt that alcohol affects the body, whether acutely or chronically. Most of us are pretty familiar with the long-term health dangers of excessive alcohol consumption, such as liver disease, pancreatitis, stroke, certain cancers, infertility, impotence, mental health problems and dementia, but let's instead focus on the potential impact of moderate drinking and short-term side-effects. Alcohol consumption is linked to around sixty diseases and these links are not just due to alcohol misuse. Moderate consumption has also been associated with an increased risk of some health harms.

For many years, it's been the belief that regularly drinking a small amount of alcohol is healthier than abstaining completely. While some research suggests that low levels of alcohol consumption may have a beneficial effect on heart health and diabetes, for instance, these findings are complex as others have found no such benefit or have concluded that other harms outweigh any potential benefits. In other words, even if it turns out that there are some small benefits, there are various potential harms of consuming any amount of alcohol so the accumulative effect may be negative overall. Scientists have also determined that low alcohol consumption is not likely to help you live longer. Reviews of lots of studies have concluded that there is no net mortality benefit of low alcohol consumption over abstention or occasional drinking. Further to this, a major study published in the *Lancet* in 2018, that considered 694 sources of data on alcohol consumption along with 592 studies on the risk of alcohol use, concluded that there is no safe level of alcohol consumption. This has been slightly disputed by at least one expert who feels that looking at global alcohol consumption and health risks is misleading as the study took into account a number of factors and health conditions that are less common in countries like the US and UK, for instance. This means that it may overestimate the harms from alcohol in societies that are healthier overall. Despite this, there is plenty of evidence that drinking any amount of alcohol is associated with an increased risk of some cancers, such as breast and oesophageal cancers, as well as cognitive decline, among other health harms. Scientific opinion is shifting away from moderate drinking as evidence suggests an overall lack of benefit, and possible harms, from even small amounts of alcohol.

Okay, so perhaps we shouldn't drink it every day but if we do decide to drink alcohol is one type of alcoholic beverage healthier than other types? Much of what I've discussed has referred to the outcomes of large studies that take into account alcohol consumption as a whole, but what about different types of drinks? Do different drinks produce different health effects? Studies assessing the effect of alcohol consumption often lump all alcoholic drinks together but this may not best reflect how people actually consume it; many drinkers confine themselves to just one or two types that they consume regularly, for instance. Is the wine drinker at the same risk of harm as the vodka drinker or beer drinker? Surely a glass of wine a day is good for us, isn't it? After all, this adage must have come from somewhere. To try and pick this apart, we need to understand what exactly is in wine and other alcoholic drinks, beyond ethanol, that may be having a physiological effect, beneficial or otherwise.

Wine

Wine is often held up as being a healthy beverage in small, regular doses. Quite a number of observational studies have found that people who regularly drink wine, red wine in particular, seem to have a lower risk of cardiovascular problems. However, these studies are not without their issues, mostly that they do not prove cause and effect, and research is ongoing to determine whether wine is really conferring a protective effect on the heart and, if so, how it is doing this. The contents of wine are an obvious starting point; what does it contain that could be having a favourable effect?

The main target of interest is polyphenols, which come from the grape skins used in wine making (this is why there are far more polyphenols in red wine than white wine as white wine isn't in contact with the skins for long). These antioxidants have been associated with a range of beneficial properties (see the *Tea* section for more on this.) Even though the bioavailability of polyphenols once consumed is low, what little is available does still exert beneficial biological effects. One type of polyphenols called flavonoids has been investigated for effects on heart health. Quercetin, in

particular, has been found to have significant antioxidant properties. Another polyphenol in wine that gets the headlines is revest-
erol. In laboratory studies, this compound has been found to exert
multiple biological effects, e.g. anti-inflammatory, antioxidant and
anti-cancer. However, while it has potential, the lab studies used
much higher doses of resveratrol than you would achieve from
drinking wine or even just eating grapes, and beneficial effects in
humans have not been demonstrated by clinical trials. Despite all
the work on individual compounds and the observational studies of
drinking habits and health, the main benefits of wine for heart
health have only so far been confirmed for older women (over age
fifty-five) and at intakes of only five units, the equivalent of around
two standard glasses of wine (175ml), *per week*. So, that's not a daily
glass after all.

Beer
What about beer? Could that be good for you? Far less research has
been published on the potential benefits of beer for health. Beer also
contains polyphenols, although fewer than wine, that could be
having a positive effect and hops themselves are used by herbalists
to treat insomnia and anxiety, but there is a lack of studies to date
demonstrating that these translate into advantageous health effects
for beer drinkers. Increased bone mineral density in male beer
drinkers was found in one study but there were anomalies in the
research underpinning this result and this effect wasn't seen in
everyone. It's also unclear whether it was the beer itself, or other
aspects of diet and lifestyle, that was responsible for this effect. A
consensus piece from a range of clinicians and researchers determined that low to moderate beer drinking reduces the risk of cardiovascular disease, but this study was funded by the Italian Association
of the Beer and Malt Industries and several of the experts declared
conflicts of interest in working for the Association or other bodies
linked to alcoholic drinks, i.e. they weren't entirely objective. At this

point in time, the jury is still out on the healthfulness of beer but there are hundreds of compounds in beer, many of which have shown promise for positive health effects in laboratory studies, and time may reveal demonstrable benefits (beyond inferred correlations) in regular consumers of this popular beverage.

Other alcoholic drinks

There is a lot less published research teasing out the health effects of other specific alcoholic drinks. Aged spirits, like whisky and cognac, contain antioxidant polyphenols, which seem to derive from wood ageing. Again, although polyphenols have been detected, we don't yet know what effect they may, or may not, be having in consumers of these drinks. Clear spirits have fewer congeners than other alcoholic drinks and are thought to be less likely to cause hangovers (I'll come onto this shortly). Some believe that gin confers health benefits, like fighting infection, relieving water retention and improving circulation, due to the juniper berries used to make it, but there are no strong scientific studies to support these claims. Yes, the berries themselves have been found to have healthful properties but we need to ask: How much of any active ingredients from the berries is actually in gin itself? How much do we need to consume to gain the benefits? How much is there in each serving of gin? Are the beneficial compounds bioavailable in humans? Etc. There are still a lot of questions to be answered before we can accurately claim that gin is healthy (as many, many lifestyle websites and publications are currently claiming).

Beyond ethanol

Evidence of a significant beneficial gain from various alcoholic drinks is still to be robustly identified but nutritional differences between them might help us to assess whether some are less harmful than others. For instance, calorie content differs widely across drinks. Here are some examples:

- Standard glass (175ml) of 12 per cent ABV wine – 126 calories
- Pint of 5 per cent ABV beer – 215 calories
- Pint of 4.5 per cent ABV cider – 277 calories
- Standard bottle (330ml) of 5 per cent ABV alcopop – 237 calories
- Single serving (25ml) of 40 per cent ABV spirit – 61 calories

If you're looking to control your calorie intake, knowing how many are in your drink might help you to choose between them. There isn't a huge amount of nutrients in alcoholic beverages, although there are slight differences in what there is. For instance, there is very little sugar in red wine (around 0.2g per 100ml) but more is found in beer (2.2g for bitter of <4 per cent ABV), medium white wine (3g in 8–13 per cent ABV wine) and sweet cider (4.3g for 3.5–5 per cent ABV cider). The quantity of some vitamins and minerals also differs, with more than three times the amount of potassium in red wine than bitter beer, for instance. Of course, some alcoholic drinks are blended with others (e.g. shandy, cocktails, rum and cola) and this will affect their nutritional content and amount of potentially beneficial compounds.

Not everyone gets them and no one yet really knows why, but some people suffer terribly. I'm describing hangovers, of course. Alcohol hangovers are generally defined as the physical and mental consequences of excessive alcohol consumption. You'll likely recognise some of the common symptoms: headache, nausea, dehydration, fatigue, weakness, dizziness, shakiness and sensitivity to light and sound. Despite their frequency and often significant impact on lives, there has actually been relatively little research done to understand what causes them, how to treat them and, crucially, how to prevent them.

Researchers have investigated a whole host of factors to see if they correlate with worse hangover symptoms to see if this will help explain the pathology behind them. Analyses of various hormones, electrolytes, free fatty acids, triglycerides, lactate, ketone bodies, cortisol, glucose and markers of dehydration have so far not found any significant association with reported alcohol hangover severity. Immune factors and differences in alcohol metabolism may play a role but a lot more work is still needed to explore these potential links. A number of other potential causes, such as sleep deprivation and smoking, have also been discounted but are acknowledged to possibly aggravate symptoms.

The internet is filled with weird and wonderful hangover cures, from bananas, raw eggs and Berocca[7] to hot baths, cabbage and ice packs (not all at once!), but unfortunately there's no really compelling scientific evidence for any of them. We're pretty familiar with the most common hangover treatments but these are really aimed at improving specific symptoms, rather than the hangover as a whole. For instance, water for the dehydration, painkillers for the headache, sugary foods for the weakness and shakiness, plain starchy foods to provide energy and settle the stomach. There's no magic bullet that will target all the symptoms, but some people may find particular things work to ease their particular symptoms.[8] Researchers are continuing to explore this area and one day someone will get very rich when they hit upon the answer. Until we really understand the pathology of the hangover, an all-encompassing, foolproof cure for it is a long way off. For now, unsurprisingly, the best way to prevent a hangover is to not drink too much in the first place, but other things you can do to try and minimise alcohol's effects is to drink water between each drink, make sure you have food in your stomach to slow down the

7 A product containing a complex of vitamins.
8 This may lead to them subsequently urging other people to take it too, which is particularly irritating when published far and wide as some sort of 'cure'. However, just because something works for one person, it doesn't mean it will work for everyone else.

absorption of the alcohol and drink water before going to bed. Nothing is guaranteed.

It might seem like a great thing to be one of those people who doesn't get hangovers, but this may not be as good as it sounds. Hangovers act as a bit of a feedback mechanism letting your body know that too much alcohol isn't good, i.e. they are something of a deterrent to future alcohol consumption. But without experiencing short-term suffering of alcohol excess (the hangover), it could lead some people to consume a higher amount of alcohol. And some researchers have suggested that this behaviour could result in alcohol use disorders in particular individuals. Other scientists have demonstrated this issue to be far more complex and that a person's hangover susceptibility may, in fact, play no such role in the risk of alcohol use disorders.

Hair of the dog

This is a term used to describe an alcoholic drink that is consumed with the purpose of lessening the effects of a hangover. According to *The Oxford English Dictionary*, the expression 'hair of the dog' stems from a longer phrase 'a hair of the dog that bites you'. It derives from an antiquated idea that if someone is bitten by a dog with rabies then they can be cured by taking a remedy containing hair from the same dog.

But does drinking alcohol help to combat a hangover? In short, no. Drinking more alcohol may make you feel better temporarily by postponing the effects of alcohol withdrawal, but it's just putting off the inevitable until a little later and you really need to let your body recover from its exposure to alcohol. (And, for those of you who are curious to know, consuming hair from a rabid dog hasn't been shown to work either.)

Beyond ethanol, other minor constituents and by-products in alcoholic drinks are often blamed for hangovers and other adverse reactions. But what's the evidence? Alcoholic drinks have also been linked to a range of short-term allergic-type reactions, including rhinitis, itching, facial swelling, headache, cough and asthma. And different effects may be associated with different drinks, such as headaches and migraines linked to red wine and heartburn linked with beer. While the causes have still to be completely pinned down, it seems that congeners may be at least partly to blame. Although ethanol is the main factor in hangovers and other unwanted short-term side-effects of alcoholic drinks, congeners have been suggested as playing a part due to their potential toxicity despite their small quantities. These minor compounds, that result from fermentation or distillation, are present in different amounts in alcoholic drinks and some research found that people who consumed drinks with a high amount (e.g. bourbon) are more likely to experience more hangovers than those who drink beverages with a much lower amount of congeners (e.g. vodka). This is disputed by other researchers who believe that congeners themselves do not lead to hangovers, although it's possible that they aggravate them.

Fining agents sometimes used in the processing of drinks, like wine, cider and beer, are also potential culprits for negative health effects. Many fining agents are derived from known allergens, such as fish, eggs and milk, and these pose an obvious risk to individuals with allergies to these. Some people are highly allergic and are likely to be aware of foods and drinks they should avoid, but many others who experience adverse reactions to such components may not realise that alcoholic drinks could pose a problem. This could be quite an issue as not all labels declare the possibility of potential allergens in their products. However, some research has found that the amount of fining agent potentially remaining in products when it reaches the consumer is very small, since most is removed by filtration during processing, and is unlikely to pose a significant risk to most people.

The gluten in the grain used to produce particular types of alcoholic drinks can be an issue for those who are intolerant of, or allergic, to gluten. This mainly refers to beers as the distillation process used to make spirits removes any traces of gluten, such as from the barley used to make malt whisky. Some manufacturers now produce specialist beers to try and avoid the problem. These are 'gluten-free' beers that have been made from ingredients that naturally don't contain gluten (e.g. rice, corn, millet, sorghum) as well as 'gluten-removed' beers that have been made with traditional ingredients but have had their gluten extracted. Gluten-free beers are more commonly available.

Some people blame their negative symptoms on sulphites, preservatives often added to alcoholic drinks to prevent spoilage, increase shelf life and reduce unwanted flavours, colours or aromas. While sensitivity to sulphites has been linked to breathing problems, such as asthma, in a few individuals, there's no good evidence to show that it is likely to cause headaches or other significant symptoms. The thing is, sulphites in alcoholic beverages are typically present in very small amounts, much lower than in other foods that you might eat that contain them, so are unlikely to be a chief culprit for ill-effects in most people. Organic wines are hailed as being free of sulphites (or at least very low in them) and so a good choice for those with wine sensitivities. However, as there's no good evidence linking sulphites to headaches and other symptoms, this may be rather a moot point.

While sulphites seem pretty benign for most, histamines in alcoholic beverages may cause health problems for some, most commonly sneezing, runny nose and sometimes stomach pain, wheezing and headache. In fact, in very high amounts these by-products of fermentation can be fatally toxic, although not usually from alcoholic drinks. Alcoholic beverages are often rich in histamines, although levels vary widely across different types (e.g. red wines contain more histamines than white wines, but champagnes also have a substantial amount). Despite this, small trials of people reporting alcohol intolerance have found no direct correlation between the histamine content of drinks and the level of intolerance experienced. So, what's going on? Researchers suggest that those with sensitivity to alcohol are likely to also have sensitivity to other foods containing histamines, such as cheese, processed meats, spinach, tomatoes, strawberries and citrus fruits. Histamines are found in a broad range of foods and it is possible that exposure to them may build up over the course of a day, for example, and alcohol then consumed in the evening might be blamed for any subsequent adverse reactions. Various makers of alcoholic beverages are making great efforts to produce drinks with lower levels of sulphites and histamines by tweaking their methods. The future will reveal whether their products lead to a decrease in reported adverse effects.

The next time someone tells you that red wine tends to give them a headache it is most likely that just having too much alcohol is the problem. Red wine contains more alcohol than white wine so sensitive types can drink less of it before suffering adverse effects. Red wine also contains more histamines than many other beverages, and if the individual is sensitive to other foods also, this could be a consideration. Something I've not covered in depth (as they largely come under congeners) is that there are also more tannins in red wine, which some blame for their headaches. There is a theory that it is linked to a sensitivity but it is an unlikely cause; after all, how many of these same people also say that they can't tolerate a strong cup of tea, which is also rich in tannins?

On a final note, there's a long-held belief that you should never mix 'grape and grain', i.e. different types of drinks, in a session because this will lead to a terrible hangover. There is no scientific evidence to show that this is true. Researchers suspect that if people are mixing drinks then they may be underestimating how much alcohol they have consumed, and the ethanol intake is more likely to cause the hangover the following morning.

Do you feel the fear?

You may find it hard to believe (or you may relate to this) but there are some people who have phobias related to alcohol. These may seem like irrational fears to some of us but for the sufferer can cause anxiety and panic and can significantly impact their day-to-day lives and relationships.

Methyphobia or potophobia (from *methy*, Greek for alcohol, or *poto*, Latin for drink) is the fear of alcohol; sufferers would avoid both drinking alcohol and those who drink it. This is often due to the fear of the physiological effects of the alcohol. Dipsophobia (from *dipso*, Greek word for thirst) is the fear of drinking alcohol. This may lead to a tendency to avoid social gathering and settings where alcohol is consumed. Zythophobia (*zythos* is the Greek word for beer) is the fear of beer. This can be due to the fear of the effects of the alcohol in beer and/or a belief that it contains live yeasts which could introduce parasites into the body. Oenophobia is the fear of wine (stemming from *oenos*, the Greek word for wine). This leads to the avoidance of wine and even of those drinking wine in case the sufferer acts unpleasantly upon the alcohol's effects.

Digestif

How we're being played (or the power of marketing)

Now that you've heard what's really behind drinks, and that some are nutritionally dubious or just plain pointless, what compels us to buy some of them? Sure, it's because there are times when we just can't face another bottle of water or need something stronger in our day, but another powerful factor is the huge marketing machine behind them. It's a big, wealthy and highly influential industry that can pour money into getting us to drink more of what they have to sell.

To be of-the-moment, our drinks need to look like a must-have accessory in one's life. Picture the celeb snapped clutching their morning coffee as they walk to their chauffeur-driven car, or a sports drink after they're seen leaving the gym. Celebrity endorsement makes a big impact. Gaining celebrity endorsement is one of the easiest and fastest ways to achieve credibility and brand recognition. When a consumer sees a celebrity they admire associated with a specific brand or product, they are prompted, whether directly or subliminally, to purchase it too. Some research has estimated that almost immediately after announcing a celebrity signing, sales of a product will likely increase by an average 4 per cent. This is before the celeb has even had to do anything.

Manufacturers don't just choose any celebrity to endorse their products; they go for those who attract their target consumer population and/or exude the image they want to associate with their products. For instance, one study published in *Pediatrics* in 2016 demonstrated that music artists popular with teenagers tend to endorse energy-dense, nutritionally poor soft drinks, as this is the key demographic for this type of beverage (e.g. colas and energy drinks). Many so-called sports drinks are fronted by well-known sportspeople, wellness drinks are linked with celebrities who are already known for their embodiment of healthy living, or at least look particularly fresh and healthy, and novel products with unusual ingredients or health claims, looking to create a new trend, pair up with social influencers and individuals who are edgy and trend-setting. The message is simple: if you drink product X, you too can be beautiful (e.g. Mila Kunis and *Jim Beam* bourbon, David Beckham and *Haig Club* whisky), strong (e.g. Arnold Schwarzenegger and *Arnold Iron Whey* protein drink), cool (e.g. Jude Law and *Johnnie Walker* whisky), sophisticated (e.g. George Clooney and *Nespresso* coffee, Roger Federer and *Jura* coffee makers) and/or on-trend (e.g. Taylor Swift and *Diet Coke*) etc. Celebrity promotion and endorsement of products leads consumers to believe that the drinks are effective, tried and tested and really do all they are claimed to do. But, at the end of the day, the celebs are there to make money and gain exposure, not support the public health of consumers.

From golfer Tiger Woods and *Gatorade* (a popular sports drink in the US), to footballer Gareth Bale and *Lucozade Sport*, to big names in pop, Beyoncé, Britney Spears, will.i.am and Justin Timberlake and *Pepsi*, the celebrity endorsements of soft drinks are endless. And manufacturers sponsor whole teams and events to attract even more followers (just think of *Red Bull* and Formula One). Industry can afford to pay big bucks to secure big names to promote their drinks and the payback is enormous – extensive fan bases mean significant markets for their products, and these endorsements translate into sales and, eventually, brand loyalty from consumers. But this all goes beyond just endorsement from celebrities. Many famous faces are heavily invested in the drinks that they are promoting. And this doesn't happen by accident. It's often the result of targeted pairing by people like Hollywood's 'Brandfather', Rohan Oza, who specialises in aligning commercial brands with the personal brands of A listers.

Madonna, a pop icon renowned for seemingly defying the ageing process and known to embrace yoga and wellness interventions, happily promotes *Vita Coco* coconut water. But then she would as she financially invested in the brand, along with other A-listers Demi Moore and Matthew McConaughey. Rihanna, another pop icon and major influencer of young people, is the face of *Vita Coco*. All of these endorsements added huge value to the product's worth and reached millions. A sports drink in the US called *Bodyarmor* received investment from basketball legend, Kobe Bryant, sending its worth soaring. This subsequently led to a host of other sporting stars lining up to endorse the drink. With millions of fans between them, you can see how their collective influence on drinking habits would be huge. Another famous soft drink investor is Jennifer Aniston, who is an advocate of *Smart Water*. This ever-youthful A-list actor, known for her devotion to healthy living, has been an investor in this company for many years and provides a huge platform for its marketing, appealing to a wide range of consumers. Hip-hop artist 50 Cent took shares in *Vitamin Water* instead of a salary after appearing in several commercials for the product. They even named a version of the drink after him, 'Formula 50'. His huge young fan base was an attractive target for the product and he continued to be associated with the drink long after he sold his shares in the company, presumably benefiting both parties.

A growing number of celebrities have even had their own drinks products over the years, including George Clooney with *Casamigos Tequila*, Jay Z with *Armand de Brignac* champagne, Dan Ackroyd with *Crystal Head* vodka, and numerous celebrities with wine blends that bear their name (e.g. Madonna, Drew Barrymore, Graham Norton, Sir Ian Botham). Kelsey Grammer has his own brewery and Sting has his own wine estate. Taylor Swift's record label, based in Tennessee, even launched its own vodka (*Big Machine Platinum Filtered Premium Vodka*). A prudent move given that the drink gains instant notoriety via its association with celebrities on the record label's books, without them even officially endorsing it.

Not so innocent?

Beyond celebrities, the drinks industry itself works hard to influence our perceptions about their products. Big industry names more frequently associated with less healthy products, like sugary soft drinks, are increasingly producing beverages to appeal to the health-conscious consumer in an effort to diversify their market. The problem is that their brands are already too well known for their classic products (e.g. Pepsi, Coca-Cola), so the healthier products need a separate identity to avoid the association with these less healthy drinks and escape consumer suspicion. This can mean that it's not always clear who the product owner really is. For example, you might be surprised to learn that the giant of smoothies, *Innocent*, renowned for its promotion of 'natural, delicious, healthy drinks that help people live well and die old',[1] and its friendly branding, is actually owned by Coca-Cola (best known for its sweetened soda drinks). The company owns around five hundred brands, such as *Rose's Lime Juice Cordial*, *Honest Coffee*, *Abbey Well* and *Dasani* bottled waters. Other companies with similarly questionable health credentials also have healthier alternatives on the market that you might not realise, such as Mars (recognised for its chocolate products), which owns *Alterra Coffee Roasters*, and PepsiCo that produces *Aquafina* water. Coca-Cola and PepsiCo control around 60 per cent of the global non-alcoholic beverages industry but as they face increasing competition from manufacturers of healthier drinks, they have been developing their own alternative beverages in response, or buying up smaller companies.

To give you a sense of just how large and powerful these companies are, seventeen of Coca-Cola's brands generate approximately $1 billion each, per year. Its advertising spend is enormous, an estimated $3.96 billion in 2017, surpassing its main competitors

---

1 Innocent's tagline. Innocent, 'Hello, we're innocent.' https://www.innocent-drinks.co.uk/us/our-story

PepsiCo and Dr Pepper. But as the taste for classic sodas declines in many developed countries, companies like Coca-Cola and PepsiCo are now investing in lower-income countries. Such countries often lack the same stringent conventions regarding advertising to children, and also face many other critical public health priorities claiming their time and attention, placing regulations surrounding sugary foodstuffs way down the agenda.

And its not just about the marketing, there's the political lobbying and research funding. The drinks industry makes sure its interests are represented at the highest levels of decision-making. Soft drinks manufacturers fought hard against plans for a sugar tax in the UK, warning that it was not the solution to reducing obesity levels. In San Francisco, the American Beverage Association created the 'Coalition for an Affordable City', supposedly a community group established to improve local matters but one that really only campaigned against a new soda tax. An investigation by *ABC News Nightline* found evidence that protesters for the group were actually recruited and paid to protest, and local businesses listed as supporters of the campaign knew nothing about their supposed involvement.[2]

Soft drinks companies, like Coca-Cola, have also been funding their own research institutes (such as the 'Beverage Institute for Health and Wellness', and the 'European Hydration Institute'), as well as sponsoring other research projects, in order to develop evidence for the benefits of their drinks, cast doubt on any harms of their drinks and promote alternative approaches (other than reducing soft drink consumption) as more effective strategies for improving people's health. They are seemingly shifting the focus onto

2 It's worth watching the ABC News Nightline report on this if you're interested: ABC News, 'When grassroots protest rallies have corporate sponsors.' https://abcnews.go.com/Nightline/video/grassroots-protest-rallies-corporate-sponsors-26671038

exercise and away from calories and sugar consumption. A lot of researchers who publish in this field may be compromised by conflicts of interest (i.e. their links with the drinks industry) (see *What are sports drinks* in Chapter 4, Cold drinks for more on this). According to an investigation published in the *British Medical Journal* in 2017, there's also evidence that soft drinks manufacturers covertly influence journalists with the message about lack of exercise being a bigger issue than sugar consumption when it comes to tackling obesity, to gain favourable media coverage. And it works. It's all about building a sense of reasonable doubt in the public's minds.

In a similar move, the alcoholic drinks industry in the UK made its views known regarding plans for minimum alcohol pricing. Already a policy in Scotland, it has been challenged in the highest courts by the alcohol industry (particularly the Scottish Whisky Association) hoping to overturn it. It emerged that representatives from the industry had numerous meetings with senior government officials in England before plans to introduce minimum alcohol pricing there were dropped, in a move described as 'deplorable' by health experts. This policy reversal went against evidence that such a minimum charge would save hundreds of lives and prevent tens of thousands of crimes each year, and it was felt by many that the Government had prioritised business interests over public health. According to an investigation by the *British Medical Journal*, similar tactics by the industry are being used to influence European alcohol policy. The issue of minimum alcohol pricing is complex and there are various arguments for and against, but it is the approach of the drinks industry that has raised concern.

Drinks industry representatives have been found to enjoy extensive access to civil servants, ministers and parliamentarians, and to invest substantial resources in fostering links with government. A research article from the London School of Hygiene and Tropical Medicine and University of York, looking at the influence of the alcohol industry on UK alcohol policy, found evidence that the industry attempts to frame issues and set agendas for policy makers to influence their outcomes. According to this research, the alcohol industry presses the case that alcohol confers a net benefit on society (e.g. via employment, tax revenues and pleasure), sponsors research favourable to their positions and supresses information that goes against their argument. And it's unlikely that it is just decision-makers in the UK that are influenced in this way.

The strategy seems to be that if you don't have a good argument for your product, then you need either to make your opponent's argument appear flawed or get people to focus on a different issue entirely (i.e. *yes, we agree there is a problem but penalising us is not the right way to go about tackling it and it won't work; you should really look at that issue over there instead*). And that is how the drinks industry is so influential. In the US, it is claimed[3] that 'Big Soda' follows the pattern set by the tobacco industry decades ago, by denying their products are unhealthy and funding research that casts doubt on scientific studies, whilst also influencing policy makers to resist taxes and regulations on the drinks industry. It is likely a similar picture in other countries.

To protect business interests, the drinks industry obviously needs to ensure that it is represented in decisions that might affect it, and make its views known. This is standard business practice and aligns with other industries that will all be trying to do the same. Where it becomes a concern is when facts are altered, suppressed or manipulated in order to achieve business gain at the cost of public health. The point here is that there are many ways to tell a tale but we have to be careful not to let a good story get in the way of hard facts. But how can we avoid the risk of being fed a load

3 Kaiser Health News. 'Soda industry steals page from tobacco to combat taxes on sugary drinks'. https://khn.org/news/soda-industry-steals-page-from-tobacco-to-combat-taxes-on-sugary-drinks/amp/?__twitter_impression=true 6 November 2018

of dubious research, washed down with a whole lot of spin? We can't all be experts and we won't always know what the truth really is, but next time you hear that there's a new study showing that drinks with a less-than-healthy image are actually good for you, or that they don't have a negative impact on health after all, ask yourself where that research has come from; who did the research and who funds the researchers? When a seemingly straightforward public health policy gets reversed or blocked, ask yourself why and in whose interests it really is.

The secret to healthy ageing? (Probably not)
It's not just celebrity endorsements and appealing branding that sell products, it's the yarn being spun that's also working its magic. Messages about what the drinks contain and what they can supposedly do for us also compel us to part with our cash. One particularly influential trend in drinks marketing is promoting the use of 'real' or 'natural' ingredients, particularly those promising health benefits. On the basis of a few basic studies and some individual anecdotes, the beliefs surrounding the benefits of wellness drinks, for example, are passed through the grapevine and amplified by the media. It doesn't take long for a blurb of scientific gobbledygook and some celebrity attention to elevate these drinks into the public consciousness, with a sense that they have been properly endorsed in some way.

Every week a new, or rediscovered ancient, breakthrough ingredient is advertised as the latest thing that will ensure our youthful longevity. It will typically promise cancer-protective, heart-healthy, cognitive-enhancing, energy-boosting, stress-busting and, of course, detoxifying qualities. You'll have seen by now that these claims are rarely justified but that doesn't deter consumers. If there's the mere possibility of a quick fix, with relatively little risk (it's mostly the bank balance that suffers), then people are willing to give products a go. But just as some ingredients launch into popularity, others fall out of favour, some merely relegated to a distant memory and others now fully maligned. Heaps of ingredients are in some way either worshipped or demonised. It's all starting to sound a bit like a cult, but in a small way it kind of is. Cults share common features, including: a) having a charismatic leader to draw in followers (popular, attractive and interesting social media influencers, endorsing particular drinks, have huge numbers of devoted 'followers'); b) spreading persuasive messaging about their beliefs and directing followers to believe (social media followers are happy to take the influencers' information at face value and support them); and c) exploiting followers for gain (social media followers will buy products endorsed by the influencer they follow as they aspire to be like them).

This isn't to say that all of these ingredients are definitely useless; it's just that we don't usually have enough clear and robust evidence of a consistent, measurable effect in humans at the amounts that we consume them. In other words, their health claims are not yet proven. In future, we may find that some of today's fashionable ingredients are in fact powerful health-promotors but until that time it's probably worth retaining a modicum of healthy scepticism.

So, is there really a 'superdrink'?

Lots of wellness drinks are marketing themselves as superdrinks but when it boils downs to it, they are nothing of the kind. Mostly a lot of hot air and unproven theories. It may sound boring but really the nearest things we have to superdrinks are arguably water and mammal milk.[4] Between them, they support all life on Earth and I think that is pretty super. Don't you?

---

4 This includes human breastmilk. Infant formula milks also do a good job but they don't contain the diverse bioactive components found in breastmilk that help protect the young baby from disease and lay down healthy foundations for their long-term development.

As with everything in this book, if you already drink particular drinks, are happy to pay potentially inflated prices, and find them helpful for you, or just like the taste, then that's marvellous. Go for it. Whatever your poison, I hope you just enjoy it. *Cheers! Chin-chin! Sláinte! Santé! Salute! Prost! Kanpie!* And jolly good health to you!

Bibliography

APERITIF

British Nutrition Foundation. 'Healthy hydration guide.' https://www.nutrition.org.uk/healthyliving/hydration/healthy-hydration-guide August 2018

British Nutrition Foundation. 'Liquids: Water.' https://www.nutrition.org.uk/nutritionscience/nutrients-food-and-ingredients/liquids.html?limit=1&start=1 July 2009

Jéquier, E. & Constant, F., 'Water as an essential nutrient: the physiological basis of hydration.' *Eur J Clin Nutr.* 2010; 64: 115–123.

1. WATERS

ANSES. Opinion of the French Agency for Food, Environmental and Occupational Health & Safety on the assessment of the safety and effectiveness of water filter jugs. 19 October 2016

Bach, C., Dauchy, X., Severin, I., Munoz, J. F., Etienne, S. & Chagnon, M. C., 'Effect of sunlight exposure on the release of intentionally and/or non-intentionally added substances from polyethylene terephthalate (PET) bottles into water: chemical analysis and in vitro toxicity.' *Food Chem.* 1 November 2014; 162: 63–71.

Bach, C., Dauchy, X., Severin, I., Munoz, J. F., Etienne, S. & Chagnon, M. C., 'Effect of temperature on the release of intentionally and non-intentionally added substances from polyethylene terephthalate (PET) bottles into water: chemical analysis and potential toxicity.' *Food Chem.* 15 August 2013; 139(1–4): 672–680.

BBC. 'Plastic particles found in bottled water.' http://www.bbc.co.uk/news/science-environment-43388870 15 March 2018

BBC *Two Trust Me I'm A Doctor.* 'Can fizzy drinks make you eat more?' http://www.bbc.co.uk/programmes/articles/29tx4RFjTKZnBsPv9R4W3DV/can-fizzy-drinks-make-you-eat-more

Beverage Marketing Corporation. 'Press Release: Bottled Water Becomes Number-One Beverage in the US.' https://www.beveragemarketing.com/news-detail.asp?id=438 10 March 2017

British Fluoridation Society. 'The extent of water fluoridation.' https://docs.wixstatic.com/ugd/014a47_0776b576cf1c49308666cef7caae934e.pdf

Brown, C. M., Dulloo, A. G. & Montani, J.-P., 'Water-Induced Thermogenesis Reconsidered: The Effects of Osmolality and Water Temperature on Energy Expenditure after Drinking.' *J Clin Endocrinol Metab*, September 2006; 91(9): 3598–3602.

Cancer Research UK. 'Do plastic bottles or food containers cause cancer?' http://www.cancerresearchuk.org/about-cancer/causes-of-cancer/cancer-controversies/plastic-bottles-and-cling-film 31 July 2018

Center For Science in The Public Interest. 'Vitaminwater Settlement Approved by Court.' https://cspinet.org/news/vitaminwater-settlement-approved-court-20160408 8 April 2016

Cheung, S. & Tai, J. 'Anti-proliferative and antioxidant properties of rosemary Rosmarinus officinalis.' *Oncol Rep*. June 2007; 17(6): 1525–1531.

Chiang, C. T., Chiu, T. W., Jong, Y. S., Chen, G. Y. & Kuo, C. D., 'The effect of ice water ingestion on autonomic modulation in healthy subjects.' *Clin Auton Res*. December 2010; 20(6): 375–380.

Coca-Cola. 'GLACEAU Smartwater.' http://www.coca-cola.co.uk/drinks/glaceau-smartwater/glaceau-smartwater

Collier, R., 'Swallowing the pharmaceutical waters.' *CMAJ*. 2012; 184(2): 163–164.

Community Preventive Services Task Force. 'Oral Health: Preventing Dental Caries, Community Water Fluoridation.' 23 January 2017

Cuomo, R., Grasso, R., Sarnelli, G., Capuano, G., Nicolai, E., Nardone, G., Pomponi, D., Budillon, G. & Ierardi, E., 'Effects of carbonated water on functional dyspepsia and constipation.' *Eur J Gastroenterol Hepatol*. September 2002; 14(9): 991-999.

Drink Water. 'Drink water is an idea.' https://www.wedrinkwater.com/pages/reason

Drinking Water Inspectorate. 'Assessment of the Effects of Jug Water Filters on the Quality of Public Water Supplies.' DWI0826 January 2003

Drinking Water Inspectorate. 'Treatment guide. Water treatment processes.' http://dwi.defra.gov.uk/private-water-supply/installations/Treatment-processes.pdf

Drinking Water Inspectorate. 'Water filters and other home treatment units.' http://dwi.defra.gov.uk/consumers/advice-leaflets/filters.pdf January 2010

Epicurious. 'The Bizarre but True Story of America's Obsession with Ice Cubes.' https://www.epicurious.com/expert-advice/why-ice-cubes-are-popular-in-america-history-freezer-frozen-tv-dinners-article 26 September 2016

EurekAlert. 'Compound found in rosemary protects against macular degeneration in laboratory model.' https://www.eurekalert.org/pub_releases/2012-11/smri-cfi112712.php 27 November 2012

Euronext. 'Naturex welcomes Rosemary extracts approval as antioxidants by the EU.' https://www.euronext.com/nl/node/294049 3 November 2010

European Federation of Bottled Waters. 'About EFBW.' http://www.efbw.org/index.php?id=24

European Food Safety Authority. 'Scientific opinion on Bisphenol A (2015).' http://www.efsa.europa.eu/sites/default/files/corporate_publications/files/factsheetbpa150121.pdf

Fenton, T. R. & Huang, T., 'Systematic review of the association between dietary acid load, alkaline water and cancer.' *BMJ Open.* 2016; 6:e010438

Food Standards Agency. 'The natural mineral water, spring water and bottled drinking water (England) regulations 2007 (As amended).' July 2010

Food Safety Magazine. 'The Sanitation of Ice-Making Equipment.' https://www.foodsafetymagazine.com/magazine-archive1/augustseptember-2013/the-sanitation-of-ice-making-equipment/ August/September 2013

Fortune. 'Coca-Cola Can't Keep Saying That VitaminWater Is Healthy.' http://fortune.com/2016/04/11/coca-cola-vitaminwater/ 11 April 2016

Francesca, N., Gaglio, R., Stucchi, C., De Martino, S., Moschetti, G. & Settanni, L., 'Yeasts and moulds contaminants of food ice cubes and their survival in different drinks.' *J Appl Microbiol.* January 2018; 124(1): 188–196.

Gaglio, R., Francesca, N., Di Gerlando, R., Mahony, J., De Martino, S., Stucchi, C., Moschetti, G. & Settanni, L., 'Enteric bacteria of food ice and their survival in alcoholic beverages and soft drinks.' *Food Microbiol.* October 2017; 67: 17–22.

Geology.com. 'Where does bottled water come from?' https://geology.com/articles/bottled-water.shtml

Gerokomou, V., Voidarou, C., Vatopoulos, A., Velonakis, E., Rozos, G., Alexopoulos, A., Plessas, S., Stavropoulou, E., Bezirtzoglou, E., Demertzis, P. G. & Akrida-Demertzi, K., 'Physical, chemical and microbiological quality of ice used to cool drinks and foods in Greece and its public health implications.' *Anaerobe.* December 2011; 17(6): 351–353.

Habtemariam, S., 'Molecular Pharmacology of Rosmarinic and Salvianolic Acids: Potential Seeds for Alzheimer's and Vascular Dementia Drugs.' *Int J Mol Sci.* February 2018; 19(2): 458.

Habtemariam, S., 'The Therapeutic Potential of Rosemary (*Rosmarinus officinalis*) Diterpenes for Alzheimer's Disease.' *Evid Based Complement Alternat Med.* 2016; 2016: 2680409.

Hampikyan, H., Bingol, E. B., Cetin, O. & Colak, H., 'Microbiological quality of ice and ice machines used in food establishments.' *J Water Health.* June 2017; 15(3): 410–417.

Harvard Health Publishing. 'What causes ice cream headache?' https://www.health.harvard.edu/pain/what-causes-ice-cream-headache 4 August 2017

Heilpflanzen-Welt Bibliothek. Commission E Monographs. 'Rosemary leaf (*Rosmarini folium*).' https://buecher.heilpflanzen-welt.de/BGA-Commission-E-Monographs/0319.htm 13 March 1990

HelpGuide. 'Vitamins and minerals. Are you getting what you need?' A Harvard Health article. https://www.helpguide.org/harvard/vitamins-and-minerals.htm

Hertin, K. J., 'A comparative study of indicator bacteria present in ice and soda from Las Vegas food establishments.' *UNLV Theses, Dissertations, Professional Papers, and Capstones.* https://digitalscholarship.unlv.edu/thesesdissertations/1282. 2011

Huck. 'Drink Water: No Fizz.' https://www.huckmag.com/perspectives/opinion-perspectives/drink-water/ 18 March 2012

Iheozor-Ejiofor, Z., Worthington, H. V., Walsh, T., O'Malley, L., Clarkson, J. E., Macey, R., Alam, R., Tugwell, P., Welch, V. & Glenny, A. M., 'Water fluoridation for the prevention of dental caries.' *Cochrane Database of Systematic Reviews.* 2015, Issue 6. Art. No.: CD010856.

Independent. 'Coca-Cola's Vitamin Drink Ad Misleading.' https://www.independent.co.uk/life-style/health-and-families/health-news/coca-colas-vitamin-drink-ad-misleading-1798719.html April 2016

John Hopkins Medicine. 'Healthy Aging. Is There Really Any Benefit to Multivitamins?' https://www.hopkinsmedicine.org/health/healthy_aging/healthy_body/is-there-really-any-benefit-to-multivitamins

Kamangar, F. & Emadi, A., 'Vitamin and mineral supplements: do we really need them?.' *Int J Prev Med.* 2012; 3(3): 221–226.

Kanduti, D., Sterbenk, P. & Artnik, B., 'Fluoride: A review of use and effects on health. *Mater Sociomed.* 2016; 28(2):133–137

Live Spring Water. 'FAQs.' https://livespringwater.com/pages/frequently-asked-questions

Mages, S., Hensel, O., Zierz, A. M., Kraya, T. & Zierz, S., 'Experimental provocation of "ice-cream headache" by ice cubes and ice water.' *Cephalalgia.* April 2017; 37(5): 464–469.

Marthaler, T. M. & Petersen, P. E., 'Salt fluoridation – an alternative in automatic prevention of dental caries.' *International Dental Journal.* 2005; 55: 351–358.

Mattsson, P., 'Headache caused by drinking cold water is common and related to active migraine.' *Cephalalgia.* April 2001; 21(3): 230–235.

Mayo Clinic. 'What is BPA, and what are the concerns about BPA?' https://www.mayoclinic.org/healthy-lifestyle/nutrition-and-healthy-eating/expert-answers/bpa/faq-20058331 11 March 2016

McArdle, W. M., Katch, F. I. & Katch, V. L., *Exercise Physiology: Nutrition, Energy, and Human Performance. Section 4: Enhancement of Energy Transfer Capacity.* Wolters Kluwer Health. 2015: 570.

Medical News Today. 'Everything you need to know about rosemary.' https://www.medicalnewstoday.com/articles/266370.php 13 December 2017

Moss, M. & Oliver, L. 'Plasma 1,8-cineole correlates with cognitive performance following exposure to rosemary essential oil aroma.' *Therapeutic Advances in Psychopharmacology.* 2012: 103–113.

Moss, M., Smith, E., Milner, M., McCreedy, J. 'Acute ingestion of rosemary water: Evidence of cognitive and cerebrovascular effects in healthy adults.' *Journal of Psychopharmacology.* 2008; 32(12): 1319–1329.

Mun, J. H. & Jun, S. S., 'Effects of carbonated water intake on constipation in elderly patients following a cerebrovascular accident.' *J Korean Acad Nurs.* April 2011; 41(2): 269–275. [Abstract only]

Naimi, M., Vlavcheski, F., Shamshoum, H. & Tsiani, E., 'Rosemary Extract as a Potential Anti-Hyperglycemic Agent: Current Evidence and Future Perspectives.' *Nutrients.* 2017; 9(9): 968.

National Health and Medical Research Council. Water Fluoridation and Human Health in Australia. NHMRC Public Statement 2017.

National Institute of Dental and Craniofacial Research. 'The Story of Fluoridation'. https://www.nidcr.nih.gov/OralHealth/Topics/Fluoride/TheStoryofFluoridation.htm July 2018

Natural Hydration Council. 'Bottled water information and FAQs.' http://www.naturalhydrationcouncil.org.uk/faqs-on-bottled-water/

NHS. 'Do I need vitamin supplements?' https://www.nhs.uk/chq/Pages/1122.aspx?CategoryID=51&SubCategoryID=168 10 October 2016

NHS Centre for Reviews and Dissemination. A Systematic Review of Water Fluoridation. Report 18. September 2000.

Nichols, G., Gillespie, I. & de Louvois, J., 'The microbiological quality of ice used to cool drinks and ready-to-eat food from retail and catering premises in the United Kingdom.' *J Food Prot.* January 2000; 63(1): 78–82.

No. 1 Rosemary Water. 'Home.' https://rosemarywater.com

NRDC. 'The Truth About Tap.' https://www.nrdc.org/stories/truth-about-tap 5 January 2016

Office of the Prime Minister's Chief Science Advisor and The Royal Society of New Zealand. 'Health effects of water fluoridation: A review of the scientific evidence. A report on behalf of the Royal Society of New Zealand and the Office of the Prime Minister's Chief Science Advisor.' August 2014

de Oliveira, D. A. & Valença, M. M., 'The characteristics of head pain in response to an experimental cold stimulus to the palate: An observational study of 414 volunteers.' *Cephalalgia*. November 2012; 32(15): 1123–1130.

Onitsuka, S., Zheng, X. & Hasegawa, H., 'Ice slurry ingestion reduces both core and facial skin temperatures in a warm environment.' *J Therm Biol*. July 2015; 51: 105–109.

Ozarowski, M., Mikolajczak, P. L., Bogacz, A., et al., '*Rosmarinus officinalis* L. leaf extract improves memory impairment and affects acetylcholinesterase and butyrylcholinesterase activities in rat brain.' *Fitoterapia*. December 2013; 91: 261–271.

Peng, C. H., Su, J. D., Chyau, C. C., Sung, T. Y., Ho, S. S., Peng, C. C. & Peng, R. Y., 'Supercritical fluid extracts of rosemary leaves exhibit potent anti-inflammation and anti-tumor effects.' *Biosci Biotechnol Biochem*. September 2007; 71(9): 2223–2232.

Perry, N. S. L., Menzies, R., Hodgson, F., Wedgewood, P., Howes, M. R., Brooker, H. J., Wesnes, K. A. & Perry, E. K., 'A randomised double-blind placebo-controlled pilot trial of a combined extract of sage, rosemary and melissa, traditional herbal medicines, on the enhancement of memory in normal healthy subjects, including influence of age.' *Phytomedicine*. 15 January 2018; 39: 42–48.

Piantadosi, C. A., ' "Oxygenated" water and athletic performance.' *Br J Sports Med*. 2006; 40(9): 740–741.

Public Health England. 'Water fluoridation: Health monitoring report for England 2014.' March 2014

Public Health England. 'Water fluoridation: Health monitoring report for England 2018.' March 2018

Rylander, R., 'Drinking water constituents and disease.' *J Nutr*. February 2008; 138(2): 423S–425S.

RxList. 'Rosemary.' https://www.rxlist.com/consumer_rosemary/drugs-condition.htm

Schoppen, S., Pérez-Granados, A. M., Carbajal, A., de la Piedra, C. & Pilar Vaquero, M. 'Bone remodelling is not affected by consumption of a sodium-rich carbonated mineral water in healthy postmenopausal women.' *Br J Nutr*. March 2005; 93(3): 339–344.

Sengupta, P., 'Potential health impacts of hard water.' *Int J Prev Med*. 2013; 4(8): 866–875.

Slate. 'The Best Water Filters.' https://slate.com/human-interest/2018/02/

the-best-water-filter-pitchers-if-youre-worried-about-lead-or-fluoride. html 20 February 2018

Snopes. 'Does "Raw Water" Provide Probiotic Benefits?' https://www.snopes. com/raw-water-provide-probiotic-health-benefits/ 11 January 2018

Spector, T., *The Diet Myth: The Real Science Behind What We Eat.* Weidenfeld & Nicolson. May 2015.

Statista. 'Bottled water consumption worldwide from 2007 to 2017 (in billion liters).' https://www.statista.com/statistics/387255/global-bottled-water-consumption/

Statista. 'Per capita consumption of bottled water worldwide in 2017, by leading countries (in gallons).' https://www.statista.com/statistics/ 183388/per-capita-consumption-of-bottled-water-worldwide-in-2009/

Summit Spring. 'Water Quality.' http://www.summitspring.com/water-quality/

The Atlantic. 'The Stubborn American Who Brought Ice to the World.' https://www.epicurious.com/expert-advice/why-ice-cubes-are-popular-in-america-history-freezer-frozen-tv-dinners-article 5 February 2013

The Chart. 'Can you explain Vitamin Water to me?' http://thechart.blogs. cnn.com/2011/04/01/can-you-explain-vitamin-water-to-me/ 1 April 2011

The Guardian. 'No ice please, we're British.' https://www.theguardian.com /lifeandstyle/2015/feb/13/iced-drinks-british-americans-ice 13 February 2015

The Mineral Calculator. 'How many minerals are in your mineral water? It's time to compare.' http://www.mineral-calculator.com/all-waters. html

The New York Times. 'Unfiltered Fervour. The Rush to Get Off the Water Grid.' https://www.nytimes.com/2017/12/29/dining/raw-water-unfiltered.html 29 December 2017

The Telegraph. 'What is raw water – and is there any sense behind Silicon Valley's latest health fad?' http://www.telegraph.co.uk/health-fitness/ body/raw-water-sense-behind-silicon-valleys-latest-health-fad/ 5 January 2018

Tourmaline Spring. 'Tourmaline Spring. Sacred Living Water.' https://tourmalinespring.com/#home

Tucker, K. L., Morita, K., Qiao, N., Hannan, M. T., Cupples, L. A. & Kiel, D. P., 'Colas, but not other carbonated beverages, are associated with low bone mineral density in older women: The Framingham Osteoporosis Study.' *Am J Clin Nutr.* October 2006; 84(4): 936–942.

Vice Sports. 'A Solution to Snowboarding's Energy Drink Problem.' https://sports.vice.com/en_us/article/yp78qy/a-solution-to-snowboardings-energy-drink-problem 15 July 2015

Wakisaka, S., Nagai, H., Mura, E., Matsumoto, T., Moritani, T. & Nagai, N., 'The effects of carbonated water upon gastric and cardiac activities and fullness in healthy young women.' *J Nutr Sci Vitaminol* (Tokyo). 2012; 58(5): 333–338.

WebMD. 'Rosemary.' https://www.webmd.com/vitamins-supplements/ingredientmono-154-rosemary.aspx?activeingredientid=154&activeingredientname=rosemary

Wikipedia on IPFS. 'Fluoridation by country.' https://ipfs.io/ipfs/QmXoypizjW3WknFiJnKLwHCnL72vedxjQkDDP1mXWo6uco/wiki/Fluoridation_by_country.html

Wired. Big Question: 'Why does tap water go stale overnight?' https://www.wired.com/2015/08/big-question-tap-water-go-stale-overnight/ 18 August 2015

World Health Organisation. 'Drinking-water.' http://www.who.int/mediacentre/factsheets/fs391/en/ 7 February 2018

World Health Organisation. 'Hardness in drinking-water. Background document for development of WHO Guidelines for Drinking-water Quality.' 2011

World Health Organisation. 'Information sheet: Pharmaceuticals in drinking-water.' http://www.who.int/water_sanitation_health/diseases-risks/risks/info_sheet_pharmaceuticals/en/

Yeo, Z. W., Fan, P. W., Nio, A. Q., Byrne, C. & Lee, J. K., 'Ice slurry on outdoor running performance in heat.' *Int J Sports Med.* November 2012; 33(11): 859–866.

Zierz, A. M., Mehl, T., Kraya, T., Wienke, A. & Zierz, S., 'Ice cream headache in students and family history of headache: a cross-sectional epidemiological study.' *J Neurol.* June 2016; 263(6): 1106–1110.

2. MILKS

Afshin, A., Micha, R., Khatibzadeh, S. & Mozaffarian, D., 'Consumption of nuts and legumes and risk of incident ischemic heart disease, stroke, and diabetes: a systematic review and meta-analysis.' *Am J Clin Nutr.* 2014; 100(1): 278–288.

Alexander, D., Bylsma, L., Vargas, A. et al. 'Dairy consumption and CVD: A systematic review and meta-analysis.' *Br J Nutr.* 2016; 115(4): 737–750.

American Society for Clinical Nutrition. 'Chapter 3. Lactose content of milk and milk products.' *Am J Clin Nutr.* 1998; 48(4): 1099–1104.

ANSC Lactation Biology Website. 'Milk composition – species table.' http://ansci.illinois.edu/static/ansc438/Milkcompsynth/milkcomp_table.html (Data adapted from: Robert D. Bremel, University of Wisconsin and from *Handbook of Milk Composition*, by R. G. Jensen, Academic Press, 1995.)

Angulo, F. J., LeJeune, J. T. & Rajala-Schultz, P. J., 'Unpasteurized Milk: A Continued Public Health Threat.' *Clinical Infectious Diseases*. 2009; 48(1): 93–100.

Atkins, P., 'School milk in Britain, 1900–1934.' *Journal of Policy History*. 2007; 19(4): 395–427.

Atkins, P., 'The milk in schools scheme, 1934–45: "nationalization" and resistance.' *History of Education*. 2001; 34: 1, 1–21.

Battelli, M. G., Polito, L. & Bolognesi, A., 'Xanthine oxidoreductase in athcrosclerosis pathogenesis: Not only oxidative stress.' *Atherosclerosis*. 2014; 237(2): 562–567.

Ballard, O. & Morrow, A. L., 'Human milk composition: nutrients and bioactive factors.' *Pediatr Clin North Am*. 2013; 60(1): 49–74.

Bath, S., Button, S. & Rayman, M., 'Iodine concentration of organic and conventional milk: Implications for iodine intake.' *Br J Nutr*. 2011; 107: 935–940.

Bath, S., Hill, S., Infante, H. G., Elghul, S., Nezianya, C. J. & Rayman, M. 'Iodine concentration of milk-alternative drinks available in the UK in comparison to cows' milk.' *Br J Nutr*. 2017; 118(7): 525–532.

Barłowska, J., Szwajkowska, M., Litwińczuk, Z. & Król, J., 'Nutritional Value and Technological Suitability of Milk from Various Animal Species Used for Dairy Production.' *Comprehensive Reviews in Food Science and Food Safety*. 2011; 10: 291–302.

BBC. 'Climate change: Which vegan milk is best?' https://www.bbc.com/news/science-environment-46654042 22 February 2019

BBC. 'The milk that lasts for months.' http://www.bbc.com/future/story/20170327-the-milk-that-lasts-forever 27 March 2017

BBC. 'Why is free milk for children such a hot topic?' https://www.bbc.co.uk/news/uk-15809645 20 November 2011

BDA. 'Food Fact Sheet: Iodine'. https://www.bda.uk.com/foodfacts/Iodine.pdf May 2016

Bee International. 'How homogenization benefits emulsions in the food industry.' http://www.beei.com/blog/how-homogenization-benefits-emulsions-in-the-food-industry 16 September 2016

Bell, S. J., Grochoski, G. T. & Clarke, A. J., 'Health Implications of Milk Containing β-Casein with the A2 Genetic Variant.' *Critical Reviews in Food Science and Nutrition*. 2006; 46:1, 93–100.

Berkeley Wellness. 'Homogenized milk myths busted.' http://www.berke-leywellness.com/healthy-eating/food/article/homogenized-milk-myths -busted 13 February 2013

Berkeley Wellness. 'Probiotics pros and cons.' http://www.berkeleywell-ness.com/supplements/other-supplements/article/probiotics-pros-and -cons 28 September 2018

BHF. 'Heart Matters. What you really need to know about milk.' https:// www.bhf.org.uk/informationsupport/heart-matters-magazine/nutri-tion/milk

Bourrie, B. C., Willing, B. P. & Cotter, P. D. 'The Microbiota and Health Promoting Characteristics of the Fermented Beverage Kefir.' *Front Microbiol.* 2016; 7: 647.

Braun-Fahrländer, C. & von Mutius, E., 'Can farm milk consumption prevent allergic diseases?' *Clin Exp Allergy.* 2011; 41(1): 29–35.

British Nutrition Foundation. 'Arsenic in rice – is it a cause for concern?' https://www.nutrition.org.uk/nutritioninthenews/headlines/arsenicin-rice.html 22 February 2017

British Nutrition Foundation. 'Dietary Fibre.' https://www.nutrition.org.uk /healthyliving/basics/fibre.html January 2018

British Nutrition Foundation. 'Saturated fat: good, bad or complex?' https:/ /www.nutrition.org.uk/nutritioninthenews/headlines/satfat.html 25 April 2017

Brooke-Taylor, S., Dwyer, K., Woodford, K. & Kost, N., 'Systematic Review of the Gastrointestinal Effects of A1 Compared with A2 β-Casein.' *Advances in Nutrition.* September 2017; 8(5): 739–748.

Centers for Disease Control and Prevention. 'Alcohol. Is it safe for mothers to breastfeed their infant if they have consumed alcohol?' https://www. cdc.gov/breastfeeding/breastfeeding-special-circumstances/vaccina-tions-medications-drugs/alcohol.html 24 January 2018

Centers for Disease Control and Prevention. 'Prescription Medication Use. Is it safe for mothers to use prescription medications while breastfeed-ing?' https://www.cdc.gov/breastfeeding/breastfeeding-special-circum-stances/vaccinations-medications-drugs/prescription-medication-use. html 24 January 2018

Centers for Disease Control and Prevention. 'Raw milk questions and answers.' https://www.cdc.gov/foodsafety/rawmilk/raw-milk-questions -and-answers.html 15 June 2017

Clemens, R. A., Hernell, O. & Michaelsen, K. F. (eds). 'Milk and Milk Products in Human Nutrition.' *Nestlé Nutr Inst Workshop Ser Pediatr Program.* 2011; 67: 187–195. Nestec Ltd., Vevey/S. Karger AG, Basel.

Clifton, P. M. et al. 'A systematic review of the effect of dietary saturated and polyunsaturated fat on heart disease. *Nutrition, Metabolism and Cardiovascular Diseases*. 2017; 27(12): 1060–1080.

CNN. 'Non-diary beverages like soy and almond milk may not be "milk", FDA suggests.' https://edition.cnn.com/2018/07/19/health/fda-soy-almond-milk-trnd/index.html 19 July 2018

Choice. 'How to buy the best milk.' https://www.choice.com.au/food-and-drink/dairy/milk/buying-guides/milk 27 April 2017

Clifford, A. J. & Swenerton, H., 'Homogenized bovine milk xanthine oxidase: A critique of the hypothesis relating to plasmalogen depletion and cardiovascular disease.' *Am J Clin Nutr.* 1983; 38(2): 327–332.

Cohen, S. M. & Ito, N., 'A Critical Review of the Toxicological Effects of Carrageenan and Processed Eucheuma Seaweed on the Gastrointestinal Tract.' *Critical Reviews in Toxicology*. 2002; 32(5): 413–444.

Dairy Council of California. 'Types of milk.' https://www.healthyeating.org/Milk-Dairy/Dairy-Facts/Types-of-Milk

DairyGood. 'Lactose-free milk: what is it and how is it made?' https://dairy-good.org/content/2014/what-is-lactose-free-milk

Dairy Processing Handbook. 'Chapter 2: The chemistry of milk.' http://dairy-processinghandbook.com/chapter/chemistry-milk 2015

Dairy Processing Handbook. 'Chapter 6.3: Homogenizers.' http://dairy-processinghandbook.com/chapter/homogenizers 2015

Dairy Processing Handbook. 'Chapter 11: Fermented milk.' http://dairy-processinghandbook.com/chapter/fermented-milk-products 2015

Dairy UK. 'Our products. Milk.' https://www.dairyuk.org/our-dairy-products/

Davis, B. J. K., Li, C. X. & Nachman, K. E., 'A Literature Review of the Risks and Benefits of Consuming Raw and Pasteurized Cow's Milk. A response to the request from The Maryland House of Delegates' Health and Government Operations Committee.' 8 December 2014

Dong, T. S. & Gupta, A., 'Influence of Early Life, Diet, and the Environment on the Microbiome.' *Clin Gastroenterol Hepatol*. 2018; 17(2): 231–242.

Eales, J., Gibson, P., Whorwell, P., et al. 'Systematic review and meta-analysis: the effects of fermented milk with Bifidobacterium lactis CNCM I-2494 and lactic acid bacteria on gastrointestinal discomfort in the general adult population.' *Therap Adv Gastroenterol*. 2016; 10(1): 74–88.

EFSA Panel on Dietetic Products, Nutrition and Allergies (NDA). 'Scientific Opinion on the substantiation of health claims related to whey protein and increase in satiety leading to a reduction in energy intake (ID 425), contribution to the maintenance or achievement of a normal body weight (ID 1683), growth and maintenance of muscle mass (ID 418, 419, 423, 426, 427, 429, 4307), increase in lean body mass during energy restriction and resistance training (ID 421), reduction of body fat mass during energy restriction and resistance training (ID 420, 421), increase in muscle strength (ID 422, 429), increase in endurance capacity during the subsequent exercise bout after strenuous exercise (ID 428), skeletal muscle tissue repair (ID 428) and faster recovery from muscle fatigue after exercise (ID 423, 428, 431), pursuant to Article 13(1) of Regulation (EC) No 1924/2006.' *EFSA Journal.* 2010; 8(10): 1818.

Evidently Cochrane. 'New Lancet Breastfeeding Series is a Call to Action.' http://www.evidentlycochrane.net/lancet-breastfeeding-series/ 29 January 2016

Financial Times. 'Dairy shows intolerance to plant-based competitors.' https://www.ft.com/content/73b37e7a-67a3-11e7-8526-7b38dcaef614 14 July 2017

Financial Times. 'Big business identifies appetite for plant-based milk.' https://www.ft.com/content/7df72c04-491a-11e6-8d68-72e9211e86ab 15 July 2016

Finglas, P.M. et al. *McCance and Widdowson's the Composition of Foods,* Seventh summary edition. Cambridge: Royal Society of Chemistry. 2015.

Food and Agriculture Organization of the United Nations. 'Dietary protein quality evaluation in human nutrition. FAO Food and Nutrition Paper 92.' 2013.

Food and Agriculture Organization of the United Nations. 'Health hazards.' http://www.fao.org/dairy-production-products/products/health-hazards/en/

Food and Agriculture Organization of the United Nations. 'Milk and milk products.' http://www.fao.org/dairy-production-products/products/en/

Food and Agriculture Organization of the United Nations. 'Milk composition.' http://www.fao.org/dairy-production-products/products/milk-composition/en/

Food and Agriculture Organization of the United Nations. 'World Milk Day: 1 June 2019.' http://www.fao.org/economic/est/est-commodities/dairy/school-milk/15th-world-milk-day/en/

Food Navigator. 'Experts make the case for European vitamin D fortification strategy.' https://www.foodnavigator.com/Article/2017/02/24/Experts-make-the-case-for-European-vitamin-D-fortification-strategy 3 April 2018

Food Navigator USA. 'Why do consumers buy plant-based dairy alternatives? And what do they think formulators need to work on?' https://www.foodnavigator-usa.com/Article/2018/02/08/Significant-percentage-of-consumers-buy-plant-based-dairy-alternatives-because-they-think-they-are-healthier-reveals-Comax-study 8 February 2018

Food Standards Agency. 'Arsenic in rice.' https://www.food.gov.uk/safety-hygiene/arsenic-in-rice 18 September 2018

Food Standards Agency. 'Raw drinking milk.' https://www.food.gov.uk/safety-hygiene/raw-drinking-milk 25 September 2018

Food Science Matters. 'What is Carrageenan?' http://www.foodsciencematters.com/carrageenan/

GOV.UK. 'Beef cattle and dairy cows: health regulations.' https://www.gov.uk/guidance/cattle-health#hormonal-treatments-and-antibiotics-for-cattle 29 August 2012

Guasch-Ferré, M., Liu, X., Malik, V. S., et al. 'Nut Consumption and Risk of Cardiovascular Disease.' *J Am Coll Cardiol.* 2017; 70(20): 2519–2532.

Guinness World Records. 'Greatest distance walked with a milk bottle balanced on the head.' http://www.guinnessworldrecords.com/world-records/greatest-distance-walked-with-a-milk-bottle-balanced-on-the-head

Guinness World Records. 'New York restaurant serves the most expensive milkshake in a glass covered with Swarovski crystals.' http://www.guinnessworldrecords.com/news/2018/6/new-york-restaurant-serves-most-expensive-milkshake-in-a-glass-covered-with-swaro-530194 20 June 2018

Hajeebhoy, N., 'Why invest, and what it will take to improve breastfeeding practices?' The Lancet Breastfeeding Series. Baby Friendly Hospital Initiative Congress. [slide set] http://www.who.int/nutrition/events/2016_bfhi_congress_presentation_latestscience_nemat.pdf 24 October 2016

Harrison, R., 'Milk Xanthine Oxidase: Hazard or Benefit?' *Journal of Nutritional & Environmental Medicine.* 2002; 12(3): 231–238.

Harvard Health. 'An update on soy: It's just so-so.' https://www.health.harvard.edu/newsletter_article/an-update-on-soy-its-just-so-so June 2010

Harvard Health. 'The hidden dangers of protein powders.' https://www.health.harvard.edu/staying-healthy/the-hidden-dangers-of-protein-powders September 2018

Harvard T. H. Chan School of Public Health. 'Straight talk about soy.' https://www.hsph.harvard.edu/nutritionsource/soy/

Heine, R. G., AlRefaee, F., Bachina, P., et al. 'Lactose intolerance and gastrointestinal cow's milk allergy in infants and children – common misconceptions revisited.' *World Allergy Organ J.* 2017; 10(1): 41.

Hill, D., Sugrue, I., Arendt, E., Hill, C., Stanton, C. & Ross, R. P., 'Recent advances in microbial fermentation for dairy and health.' *F1000Res.* 2017; 6: 751.

Ho, J., Maradiaga, I., Martin, J., Nguyen, H. & Trinh, L., 'Almond milk vs. cow milk life cycle assessment.' http://www.environment.ucla.edu/perch/resources/images/cow-vs-almond-milk-1.pdf 2 June 2016

Infant Nutrition Council, Australia & New Zealand. 'Breastmilk Information.' http://www.infantnutritioncouncil.com/resources/breastmilk-information/

Institute of Medicine (US) Committee on the Evaluation of the Addition of Ingredients New to Infant Formula. *Infant Formula: Evaluating the Safety of New Ingredients.* Washington (DC): National Academies Press (US); 2004. 3, Comparing Infant Formulas with Human Milk. https://www.ncbi.nlm.nih.gov/books/NBK215837/

Jenness, R., 'The composition of human milk.' *Semin Perinatol.* July 1979; 3(3): 225–239.

Kroger, M., Kurmann, J. A. & Rasic, J. L., 'Fermented milks: past, present, and future.' *Food Technology.* 1989: 43: 92–99.

Kwok, T. C., Ojha, S. & Dorling, J., 'Feed thickeners in gastro-oesophageal reflux in infants.' *BMJ Paediatrics Open.* 2018; 2: e000262.

Lacroix, M., Bon, C., Bos, C., Léonil, J., Benamouzig, R., Luengo, C., Fauquant, J., Tomé, D. & Gaudichon, C., 'Ultra High Temperature Treatment, but Not Pasteurization, Affects the Postprandial Kinetics of Milk Proteins in Humans.' *The Journal of Nutrition.* 2008; 138(12): 2342–2347.

Lawrance, P., 'An Evaluation of Procedures for the Determination of Folic acid in Food by HPLC. A Government Chemist Programme Report [No. LGC/R/2011/180].' September 2011.

Li, X., Meng, X., Gao, X., et al. 'Elevated Serum Xanthine Oxidase Activity Is Associated With the Development of Type 2 Diabetes: A Prospective Cohort Study.' *Diabetes Care Apr.* 2018; 41(4): 884–890.

Lordan, R., Tsoupras, A., Mitra, B. & Zabetakis, I., 'Dairy Fats and Cardiovascular Disease: Do We Really Need to be Concerned?' *Foods.* 2018;7(3): 29.

Lucey, J. A., 'Raw Milk Consumption: Risks and Benefits.' *Nutr Today.* 2015; 50(4): 189–193.

Macdonald, L. E., Brett, J., Kelton, D., Majowicz, S. E., Snedeker, K. & Sargeant, J. M., 'A systematic review and meta-analysis of the effects of pasteurization on milk vitamins, and evidence for raw milk consumption and other health-related outcomes.' *J Food Prot.* 2011; 74(11): 1814–1832.

Mäkinen, O. E., Wanhalinna, V., Zannini, E. & Arendt, E. K., 'Foods for Special Dietary Needs: Non-dairy Plant-based Milk Substitutes and Fermented Dairy-type Products.' *Critical Reviews in Food Science and Nutrition.* 2016; 56:3, 339–349.

Manners, J. & Craven, H., 'Milk: Processing of Liquid Milk.' *Encyclopedia of Food Sciences and Nutrition* (Second Edition). 2003.

Marangoni, F., Pellegrino, L., Verduci, E. et al. 'Cow's Milk Consumption and Health: A Health Professional's Guide.' *J Am Coll Nutr.* March–April 2019; 38(3): 197–208.

Market Screener. 'A2 Milk: Controversial New Milk Shakes Up Big Dairy.' https://www.marketscreener.com/A2-MILK-COMPANY-LTD-21453329 /news/A2-Milk-Controversial-New-Milk-Shakes-Up-Big-Dairy-26416832/ 24 April 2018

Martin, C. R., Ling, P. R. & Blackburn, G. L., 'Review of Infant Feeding: Key Features of Breast Milk and Infant Formula.' *Nutrients.* 2016; 8(5): 279.

Mayo Clinic. 'Milk allergy.' https://www.mayoclinic.org/diseases-conditions/milk-allergy/symptoms-causes/syc-20375101 6 June 2016

McGill. 'Battle of the milks: Are plant-based milks appropriate for children?' https://www.mcgill.ca/oss/article/health-and-nutrition/battle-milks-are-plant-based-milks-appropriate-children 16 November 2017

McKevith, B. & Shortt, C., 'Fermented Milks: Other Relevant Products.' *Encyclopedia of Food Sciences and Nutrition* (Second Edition). 2003: 2383–2389.

Meharg, A. A., Deacon, C., Campbell, R. C. J., Carey, A. M., Williams, P. N., Feldmann, J. & Raab, A., 'Inorganic arsenic levels in rice milk exceed EU and US drinking water standards.' *J. Environ. Monit.* 2008; 10: 428–431.

Meléndez-Illanes, L., González-Díaz, C., Chilet-Rosell, E. & Álvarez-Dardet, C., 'Does the scientific evidence support the advertising claims made for products containing *Lactobacillus casei* and *Bifidobacterium lactis*? A systematic review.' *Journal of Public Health.* 2016; 38(3): e375–e383.

Michalski, M. C., 'On the supposed influence of milk homogenization on the risk of CVD, diabetes and allergy.' *Br J Nutr.* 2007; 97(4): 598–610.

Mills Oakley. 'Hemp-based foods.' https://www.millsoakley.com.au/hemp -based-foods-to-be-legalised-in-australia/ May 2017

Mintel. 'US non-dairy milk sales grow 61% over the last five years.' http:// www.mintel.com/press-centre/food-and-drink/us-non-dairy-milk-sales -grow-61-over-the-last-five-years 4 January 2018

Morton, R. W., Murphy, K. T., McKellar, S. R., et al. 'A systematic review, meta-analysis and meta-regression of the effect of protein supplementation on resistance training-induced gains in muscle mass and strength in healthy adults.' *Br J Sports Med.* 2017; 52(6): 376–384.

National Dairy Council. 'Understanding the science behind A2 milk.' https://www.nationaldairycouncil.org/content/2015/understanding-the-science-behind-a2-milk 8 February 2017

National Institute on Alcohol Abuse and Alcoholism. 'Alcohol's Effect on Lactation.' https://pubs.niaaa.nih.gov/publications/arh25-3/230-234.htm

Nature. 'Archaeology: The milk revolution.' https://www.nature.com/news/archaeology-the-milk-revolution-1.13471 31 July 2013

NCT. 'Formula feeding: what's in infant formula milk?' https://www.nct.org.uk/baby-toddler/feeding/early-days/formula-feeding-whats-infant-formula-milk

New Scientist. 'Probiotics are mostly useless and can actually hurt you.' https://www.newscientist.com/article/2178860-probiotics-are-mostly-useless-and-can-actually-hurt-you/ 6 September 2018

Newcastle University. 'Study finds clear differences between organic and non-organic products.' https://www.ncl.ac.uk/press/articles/archive/2016/02/organicandnon-organicmilkandmeat/ 16 February 2016

NICE. 'Postnatal care. Quality statement 6: Formula feeding. Quality standard [QS37].' Published date: July 2013. Last updated: June 2015.

NICE Clinical Knowledge Summaries. 'GORD in children.' https://cks.nice.org.uk/gord-in-children March 2015

Nieminen, M. T., Novak-Frazer, L., Collins, R., et al. 'Alcohol and acetaldehyde in African fermented milk mursik – a possible etiologic factor for high incidence of esophageal cancer in western Kenya.' *Cancer Epidemiol Biomarkers Prev.* 2012; 22(1): 69–75.

NIH Genetics Home Reference. 'Lactose intolerance.' https://ghr.nlm.nih.gov/condition/lactose-intolerance#statistics

NHS. 'Lactose intolerance. Causes.' https://www.nhs.uk/conditions/lactose-intolerance/causes/ 25 February 2019

NHS. 'What should I do if I think my baby is allergic or intolerant to cows' milk?' https://www.nhs.uk/common-health-questions/childrens-health/what-should-i-do-if-i-think-my-baby-is-allergic-or-intolerant-to-cows-milk/ 13 July 2019

Ocado. 'Essential Waitrose Longlife Unsweetened Soya Drink 1Ltr.' https://www.ocado.com/webshop/product/Essential-Waitrose-Longlife-Unsweetened-Soya-Drink/14031011

OECD iLibrary. 'OECD-FAO Agricultural Outlook 2017–2026.' https://

www.oecd-ilibrary.org/docserver/agr_outlook-2017-en.pdf? 10 July 2017

Ojo-Okunola, A., Nicol, M. & Du Toit, E., 'Human Breast Milk Bacteriome in Health and Disease.' *Nutrients*. 2018; 10: 1643.

Oliver, S. P., Boor, K. J., Murphy, S. C. & Murinda. S. E., 'Food Safety Hazards Associated with Consumption of Raw Milk.' *Foodborne Pathogens and Disease*. 2009; 6(7): 793–806.

Önning, G., Wallmark, A., Persson, M., Åkesson, B., Elmståhl, S. & Öste, R., 'Consumption of Oat Milk for 5 Weeks Lowers Serum Cholesterol and LDL Cholesterol in Free-Living Men with Moderate Hypercholesterolemia.' *Ann Nutr Metab*. 1999; 43: 301–309.

Ontario Public Health Association. 'Balancing and communication issues related to environmental contaminants in breastmilk.' http://www.opha.on.ca/OPHA/media/Resources/Resource%20Documents/2004-01_pp.pdf?ext=.pdf March 2004

Ottaway, P. B., 'The stability of vitamins in fortified foods and supplements.' *Food Fortification and Supplementation: Technological, Safety and Regulatory Aspects*. Woodhead Publishing. 2008.

Pal, S., Woodford, K., Kukuljan, S. & Ho, S., 'Milk Intolerance, Beta-Casein and Lactose.' *Nutrients*. 2015; 7(9): 7285–7297.

Pimenta, F. S., Luaces-Regueira, M., Ton, A. M. M., Campagnaro, B. P., Campos-Toimil, M., Pereira, T. M. C. & Vasquez, E. C., 'Mechanisms of Action of Kefir in Chronic Cardiovascular and Metabolic Diseases.' *Cell Physiol Biochem*. 2018; 48: 1901–1914.

Plant Based News. 'Global plant milk market set to top a staggering $16 billion in 2018.' https://www.plantbasednews.org/post/global-plant-milk-market-set-to-top-a-staggering-16-billion-in-2018 15 June 2017

Plant Based News. 'UK milk alternative sector to soar by 43% over next four years.' https://www.plantbasednews.org/post/uk-milk-alternative-sector-to-soar-by-43-by-2022 7 December 2017

Quartz. 'Ten years after China's infant milk tragedy, parents still won't trust their babies to local formula.' https://qz.com/1323471/ten-years-after-chinas-melamine-laced-infant-milk-tragedy-deep-distrust-remains/ 16 July 2018

Quartz. 'There's a war over the definition of "milk" between dairy farmers and food startups – and Trump may settle it.' https://qz.com/923234/theres-a-war-over-the-definition-of-milk-between-dairy-farmers-and-food-startups-and-donald-trump-may-settle-it/ 4 March 2017

Quigley, L., O'Sullivan, O., Stanton, C., Beresford, T. P., Ross, R. P., Fitzgerald, G. F. & Cotter, P. D., 'The complex microbiota of raw milk.' *FEMS Microbiology Reviews*. 2013; 37(5): 664–698.

Rautava, S., 'Early microbial contact, the breast milk microbiome and child health.' *J Dev Orig Health Dis*. February 2016; 7(1): 5–14.

RCPCH. 'Position statement: breastfeeding in the UK.' https://www.rcpch.ac.uk/resources/position-statement-breastfeeding-uk 3 May 2018

Reuters. 'Competition heats up for controversial a2 Milk Company.' https://uk.reuters.com/article/us-a2-milk-company-strategy-analysis/competition-heats-up-for-controversial-a2-milk-company-idUKKCN1IH0T9 16 May 2018

Reuters. 'French prosecutors step up probe into baby milk contamination at Lactalis.' https://uk.reuters.com/article/us-france-babymilk-investigation/french-prosecutors-step-up-probe-into-baby-milk-contamination-at-lactalis-idUKKCN1MJ1RR 10 October 2018

Ripple. 'Original Nutritious Pea Milk.' https://www.ripplefoods.com/original-plant-milk/

Rosa, D. D., Dias, M. M. S., Grześkowiak, L. M. et al. 'Milk kefir: nutritional, microbiological and health benefits'. *Nutr Res Rev*. 2017; 30 (1): 82–96.

RTRS. 'Mission and vision.' http://www.responsiblesoy.org/about-rtrs/mission-and-vision/?lang=en

Sachs, H. C., Committee on Drugs. 'The Transfer of Drugs and Therapeutics into Human Breast Milk: An Update on Selected Topics.' *Pediatrics*. September 2013; 132(3): e796-e809.

Sainsbury's. 'Alpro Hazelnut UHT Drink.' https://www.sainsburys.co.uk/shop/gb/groceries/dairy-free-drinks-/alpro-long-life-milk-alternative--hazelnut-1l

Sainsbury's. 'Alpro Roasted Almond Milk Original UHT Drink.' https://www.sainsburys.co.uk/shop/gb/groceries/dairy-free-drinks-/alpro-long-life-almond-milk-alternative-1l

Sainsbury's. 'Innocent Almond Dairy Free 750ml.' https://www.sainsburys.co.uk/shop/ProductDisplay

Sainsbury's. 'Rude Health Almond Drink.' https://www.sainsburys.co.uk/shop/gb/groceries/dairy-free-drinks-/rude-health-uht-almond-milk-1l

Science Daily. 'Further knowledge required about the differences between milk proteins.' https://www.sciencedaily.com/releases/2017/04/170428102103.htm 28 April 2017

Science Daily. ' "Organic milk" is poorer in iodine than conventional milk.' https://www.sciencedaily.com/releases/2013/07/130704094630.htm 4 July 2013

Science Media Centre. 'Expert reaction to differences between organic and conventional milk and meat.' http://www.sciencemediacentre.org/expert-reaction-to-differences-between-organic-and-conventional-milk-and-meat/ 16 February 2016

Scott, K. J. & Bishop, D. R., 'Nutrient content of milk and milk products: vitamins of the B complex and vitamin C in retail market milk and milk products.' *International Journal of Dairy Technology*. 1986; 39: 32–35.

Sethi, S., Tyagi, S. K. & Anurag, R. K., 'Plant-based milk alternatives an emerging segment of functional beverages: a review.' *J Food Sci Technol*. 2016; 53(9):3408–3423.

Silanikove, N., Leitner, G. & Merin, U., 'The Interrelationships between Lactose Intolerance and the Modern Dairy Industry: Global Perspectives in Evolutional and Historical Backgrounds.' *Nutrients*. 2015; 7: 7312–7331.

Soil Association. 'Antibiotic use in dairy and beef farming.' https://www.soilassociation.org/our-campaigns/save-our-antibiotics/reduce-antibiotics-use-on-your-farm/cows/

Soil Association. 'Organic beef and dairy cows.' https://www.soilassociation.org/organic-living/whyorganic/better-for-animals/organic-cows/

Soil Association. 'Organic milk – more of the good stuff!' https://www.soilassociation.org/blogs/2017/organic-milk-more-of-the-good-stuff/ 6 April 2017

Sousa, A. & Kopf-Bolanz, K. A., 'Nutritional Implications of an Increasing Consumption of Non-Dairy Plant-Based Beverages Instead of Cow's Milk in Switzerland.' *J Adv Dairy Res*. 2017; 5: 197.

Soyinfo Center. 'History of soymilk and dairy-like soymilk products.' http://www.soyinfocenter.com/HSS/soymilk1.php 2004

St-Onge, M. P., Mikic, A. & Pietrolungo, C. E., 'Effects of Diet on Sleep Quality.' *Adv Nutr*. 2016; 7(5): 938–949.

Statista. 'Annual consumption of fluid cow milk worldwide in 2018, by country (in 1,000 metric tons).' https://www.statista.com/statistics/272003/global-annual-consumption-of-milk-by-region/

Statista. 'Per capita consumption of fluid milk worldwide in 2016, by country (in liters).' https://www.statista.com/statistics/535806/consumption-of-fluid-milk-per-capita-worldwide-country/

Stobaugh, H., 'Maximizing Recovery and Growth When Treating Moderate Acute Malnutrition with Whey-Containing Supplements.' *Food and Nutrition Bulletin*. 2018; 39(2 Suppl): S30–S34.

Stobaugh, H. C., Ryan, K. N., Kennedy, J. A., Grise, J. B., Crocker, A. H., Thakwalakwa, C., Litkowski, P. E., Maleta, K. M., Manary, M. J. & Trehan, I., 'Including whey protein and whey permeate in ready-to-use supplementary food improves recovery rates in children with moderate acute malnutrition: a randomized, double-blind clinical trial.' *Am J Clin Nutr*. 2016; 103(3): 926–933.

Sustainable Food Trust. 'Milk: The sustainability issue.' https://sustainablefoodtrust.org/articles/milk-the-sustainability-issue/ 12 January 2017

Szajewska, H. & Shamir, R. (eds), 'Evidence-Based Research in Pediatric Nutrition.' *World Rev Nutr Diet*. Basel, Karger, 2013; 108: 56–62.

Tam, H. K., Kelly, A. S., Metzig, A. M., Steinberger, J. & Johnson, L. A., 'Xanthine oxidase and cardiovascular risk in obese children.' *Child Obes*. 2014; 10(2): 175–180.

Tamime, A. Y., 'Fermented milks: a historical food with modern applications – a review.' *Eur J Clin Nutr*. 2002; 56(Suppl 4): S2–S15.

TES. 'Return of free milk in schools to be "considered" by government.' https://www.tes.com/news/return-free-milk-schools-be-considered-government 11 February 2016

Tesco. 'Alpro Coconut Fresh Drink.' https://www.tesco.com/groceries/en-GB/products/282925010

Tesco. 'Alpro Soya Longlife Drink Alternative 1 Litre.' https://www.tesco.com/groceries/en-GB/products/251523947

Tesco. 'Koko Dairy Free Original Plus Calcium Drink Alternative.' https://www.tesco.com/groceries/en-GB/products/276993737

Tesco. 'Tesco Longlife Soya Drink Sweetened 1Ltr.' https://www.tesco.com/groceries/en-GB/products/256438810

The Dairy Council. 'Milk factsheet.' https://www.milk.co.uk/hcp/wp-content/uploads/sites/2/woocommerce_uploads/2016/12/Milk_consumer_2016.pdf 2016

The Dairy Council. 'Milk. Nutrition information for all the family.' https://www.milk.co.uk/hcp/wp-content/uploads/sites/2/woocommerce_uploads/2016/12/Milk-Consumer-2018.pdf

The Guardian. 'Lactalis to withdraw 12m boxes of baby milk in salmonella scandal.' https://www.theguardian.com/world/2018/jan/14/lactalis-baby-milk-salmonella-scandal-affects-83-countries-ceo-says 15 January 2018

The Guardian. 'Avoiding meat and dairy is "single biggest way" to reduce your impact on Earth' https://www.theguardian.com/environment/2018/may/31/avoiding-meat-and-dairy-is-single-biggest-way-to-reduce-your-impact-on-earth 1 June 2018

The Grocer. 'Asda boosts fortified milk lineup with own label and branded lines.' https://www.thegrocer.co.uk/buying-and-supplying/new-product-development/asda-adds-own-label-and-branded-fortified-milk-lines/562166.article 15 January 2018

The Grocer. 'UK milk sales down £240m over two years.' https://www.thegrocer.co.uk/buying-and-supplying/categories/dairy/uk-milk-sales-down-240m-over-two-years/546272.article 16 December 2016

The Lancet. 'Web appendix 4: Lancet breastfeeding series paper 1. data sources and estimates: countries without standardized surveys.' 2016. www.thelancet.com/cms/attachment/2047468706/2057986218 /mmc1.pdf

The Telegraph. 'Farmers say non-dairy should not be described as "milk".' https://www.telegraph.co.uk/news/2017/06/09/nfu-says-non-dairy-should-not-described-milk/ 9 June 2017

The Telegraph. 'Government to consider bringing back free milk in schools to boost children's health.' https://www.telegraph.co.uk/education/ 12152492/Government-to-consider-bringing-back-free-milk-in-schools-to-boost-childrens-health.html 11 February 2016

Thorning, T. K., Raben, A., Tholstrup, T., Soedamah-Muthu, S. S., Givens, I. & Astrup, A., 'Milk and dairy products: good or bad for human health? An assessment of the totality of scientific evidence.' Food Nutr Res. 2016; 60: 32527.

Truswell, A. S., 'The A2 milk case: a critical review.' Eur J Clin Nutr. 2005; 59(5): 623–631.

Turck, D., 'Cow's milk and goat's milk.' World Rev Nutr Diet. 2013; 108: 56–62.

Unicef. 'Breastfeeding. A Mother's Gift, for Every Child.' https://www. unicef.org/publications/files/UNICEF_Breastfeeding_A_Mothers_ Gift_for_Every_Child.pdf 2018

Unicef. 'A guide to infant formula for parents who are bottle feeding: The health professionals' guide.' https://www.unicef.org.uk/babyfriendly/ wp-content/uploads/sites/2/2016/12/Health-professionals-guide-to-infant-formula.pdf 2014

University of Guelph. 'Homogenization of mix.' https://www.uoguelph.ca /foodscience/book-page/homogenization-mix

University of Guelph. 'Pathogenic microorganisms in milk.' https://www. uoguelph.ca/foodscience/book-page/pathogenic-microorganisms-milk

Vanga, S. K. & Raghavan, V., 'How well do plant based alternatives fare nutritionally compared to cow's milk?' J Food Sci Technol. 2018; 55 (1): 10–20.

Victoria, C. G., Bahl, R., Barros, A. J. D., et al, for The Lancet Breastfeeding Series Group. 'Breastfeeding in the 21st century: epidemiology, mechanisms, and lifelong effect.' Lancet 2016; 387(10017): 457–490.

Vojdani, A., Turnpaugh, C. & Vojdani, E., 'Immune reactivity against a variety of mammalian milks and plant-based milk substitutes.' Journal of Dairy Research. 2018; 85(3): 358–365.

Vox. ' "Fake milk": why the dairy industry is boiling over plant-based milks.' https://www.vox.com/2018/8/31/17760738/almond-milk-dairy-soy-oat-labeling-fda 21 December 2018

Waitrose. 'Alpro Chilled Oat.' https://www.waitrose.com/ecom/products/alpro-chilled-oat/689817-558475-558476

Waitrose. 'Alpro Longlife Original Rice Drink.' https://www.waitrose.com/ecom/products/alpro-longlife-original-rice-drink/757659-260066-260067

Waitrose. 'Good Hemp Longlife Alternative to Milk.' https://www.waitrose.com/ecom/products/good-hemp-longlife-alternative-to-milk/370422-52082-52083

Waitrose. 'Innocent Oat Dairy Free.' https://www.waitrose.com/ecom/products/innocent-oat-dairy-free/437292-659568-659569

Waitrose. 'OOO Mega Plantbased Flax Drink.' https://www.waitrose.com/ecom/products/ooo-mega-plantbased-flax-drink/717656-601770-601771

Waitrose. 'Responsible soya.' https://www.waitrose.com/home/inspiration/about_waitrose/the_waitrose_way/responsible-soya-sourcing.html

Waitrose. 'Rude Health Organic Longlife Brown Rice Drink.' https://www.waitrose.com/ecom/products/rude-health-organic-longlife-brown-rice-drink/822624-287775-287776

Waitrose. 'Vita Coco Coconut Milk Original.' https://www.waitrose.com/ecom/products/vita-coco-coconut-milk-original/827516-668776-668777

Witard, O. C., Jackman, S. R., Breen, L., Smith, K., Selby, A. & Tipton, K. D., 'Myofibrillar muscle protein synthesis rates subsequent to a meal in response to increasing doses of whey protein at rest and after resistance exercise.' *Am J Clin Nutr.* 2014; 99 (1): 86–95.

World Cancer Research Fund. 'Could soya products affect my risk of breast cancer?' https://www.wcrf-uk.org/uk/blog/articles/2017/10/could-soya-products-affect-my-risk-breast-cancer 19 October 2017

World Health Organisation. 'Breastfeeding.' http://www.who.int/topics/breastfeeding/en/

WWF. 'Dairy. Overview.' https://www.worldwildlife.org/industries/dairy

WWF. 'Soy. Overview.' https://www.worldwildlife.org/industries/soy

Which? 'Choosing the right formula milk. Toddler formula milk.' https://www.which.co.uk/reviews/formula-milk/article/choosing-the-right-formula-milk/toddler-formula-milk

World Cancer Research Fund. 'Meat, fish and dairy products and the risk of cancer.' https://www.wcrf.org/sites/default/files/Meat-Fish-and-Dairy-products.pdf 2018

Zamora, A., Ferragut, V., Guamis, B. & Trujillo, A. J., 'Changes in the surface protein of the fat globules during ultra-high pressure homogenisation and conventional treatments of milk.' *Food Hydrocolloids.* 2012; 29(1): 135–143.

3. HOT DRINKS

American Cancer Society. 'World Health Organization Says Very Hot Drinks May Cause Cancer.' https://www.cancer.org/latest-news/world-health-organization-says-very-hot-drinks-may-cause-cancer.html June 15 2016

American Museum of Tort Law. 'Liebeck v. McDonald's.' https://www.tort-museum.org/liebeck-v-mcdonalds/ 13 June 2016

Andrici, J. & Eslick, G. D., 'Hot Food and Beverage Consumption and the Risk of Esophageal Cancer: A Meta-Analysis.' *Am J Prev Med.* 2015; 49(6): 952–960.

Anila Namboodiripad, P. & Kori, S., 'Can coffee prevent caries?' *J Conserv Dent.* 2009; 12(1): 17–21.

Arab, L., Khan, F. & Lam, H., 'Epidemiologic evidence of a relationship between tea, coffee, or caffeine consumption and cognitive decline.' *Adv Nutr.* 2013; 4(1): 115–122.

Araújo, L. F., Mirza, S. S., Bos, D., Niessen, W. J., Barreto, S. M., van der Lugt, A., Vernooij, M. W., Hofman, A., Tiemeier, H. & Ikram, M. A., 'Association of Coffee Consumption with MRI Markers and Cognitive Function: A Population-Based Study.' *J Alzheimers Dis.* 2016; 53(2): 451–461.

Araújo, L. F., Giatti, L., Reis, R. C. P., Goulart, A. C., Schmidt, M. I., Duncan, B. B., Ikram, M. A. & Barreto, S. M., 'Inconsistency of Association between Coffee Consumption and Cognitive Function in Adults and Elderly in a Cross-Sectional Study (ELSA-Brasil). *Nutrients.* 2015; 7: 9590–9601.

Australian Bureau of Statistics. ' "Caffeine" Australian Health Survey: Usual Nutrient Intakes, 2011–12.' 6 March 2015.

Bae, J. H., Park, J. H., Im, S. S. & Song, D. K., 'Coffee and health.' *Integr Med Res.* 2014; 3(4): 189–191.

Bain, A. R., Lesperance, N. C. & Jay, O. 'Body heat storage during physical activity is lower with hot fluid ingestion under conditions that permit full evaporation.' *Acta Physiol (Oxf).* October 2012; 206(2): 98–108.

BBC. 'PG Tips to switch to plastic-free teabags.' http://www.bbc.co.uk/news/uk-43224797 28 February 2018

BBC. 'The food supplement that ruined my liver.' https://www.bbc.co.uk/news/stories-45971416 25 October 2018

BBC. 'Coffee: Who grows, drinks and pays the most?' http://www.bbc.co.uk/news/business-43742686 13 April 2018

Beverage Daily.com. 'The 42 Degrees Company launches self-heating coffee-cans.' https://www.beveragedaily.com/Article/2018/08/27/The-42-Degrees-Company-launches-self-heating-coffee-cans. 27 August 2018

Boyle, N. B., Lawton, C. & Dye, L., 'The Effects of Magnesium Supplementation on Subjective Anxiety and Stress-A Systematic Review.' *Nutrients.* 2017; 9(5): 429.

Bracesco, N., Sanchez, A. G., Contreras, V., Menini, T. & Gugliucci, A., 'Recent advances on Ilex paraguariensis research: minireview.' *J Ethnopharmacol.* 2011; 136(3): 378–384.

British Coffee Association 'Coffee in the UK.' http://www.britishcoffeeassociation.org/about_coffee/from_bean_to_cup/decaffeination/

British Nutrition Foundation. 'Pregnancy and pre-conception.' https://www.nutrition.org.uk/nutritionscience/life/pregnancy-and-pre-conception.html?showall=1 January 2016

Brown, F. & Diller, K. R., 'Calculating the optimum temperature for serving hot beverages.' *Burns.* 2008; 34(5): 648–654.

Brzezicha-Cirocka, J., Grembecka, M. & Szefer, P., 'Monitoring of essential and heavy metals in green tea from different geographical origins.' *Environ Monit Assess.* 2016; 188(3): 183.

Cabrera, C., Artacho, R. & Giménez, R., 'Beneficial effects of green tea – a review.' *J Am Coll Nutr.* 2006; 25(2): 79–99.

Cadbury. 'Hot Chocolate Instant.' https://www.cadbury.co.uk/products/cadbury-hot-chocolate-instant-11688

Campaign 'Nescafe discards self-heating cans.' https://www.campaignlive.co.uk/article/nescafe-discards-self-heating-cans/155450?src_site=marketingmagazine 14 August 2002

Carman, A. J., Dacks, P. A., Lane, R. F., Shineman, D. W. & Fillit, H. M., 'Current evidence for the use of coffee and caffeine to prevent age-related cognitive decline and Alzheimer's disease.' *J Nutr Health Aging.* 2014; 18(4): 383–392.

Chacko, S. M., Thambi, P. T., Kuttan, R. & Nishigaki, I., 'Beneficial effects of green tea: a literature review.' *Chin Med.* 2010; 5: 13.

Chemistry World. 'The chemistry in your cuppa.' https://www.chemistryworld.com/feature/the-chemistry-in-your-cuppa/2500010.article 5 December 2016

Chemistry World. 'Uncovering the secrets of tea.' https://www.chemistryworld.com/news/uncovering-the-secrets-of-tea/5634.article

Chemistry World. 'Chemistry in every cup.' https://www.chemistryworld.com/feature/chemistry-in-every-cup/3004537.article

Chu, D. C. & Juneja, L. R., 'General chemical composition of green tea and its infusion.' In Yamamoto, T., Juneja, L. R., Chu, D. C. & Kim, M. (eds.), *Chemistry and Applications of Green Tea.* CRC Press, Boca Raton. 1997: 13–22.

Chung, K. T., Wong, T. Y., Wei, C. I., Huang, Y. W. & Lin, Y., 'Tannins and human health: a review.' *Crit Rev Food Sci Nutr.* 1998; 38(6): 421–464.

Church, D. D., Hoffman, J. R., LaMonica, M. B., Riffe, J. J., Hoffman, M. W., Baker, K. M., Varanoske, A. N., Wells, A. J., Fukuda, D. H. & Stout, J. R., 'The effect of an acute ingestion of Turkish coffee on reaction time and time trial performance.' *J Int Soc Sports Nutr.* 2015; 12: 37.

Cleverdon, R., Elhalaby, Y., McAlpine, M. D., Gittings, W. & Ward, W. E., 'Total Polyphenol Content and Antioxidant Capacity of Tea Bags: Comparison of Black, Green, Red Rooibos, Chamomile and Peppermint over Different Steep Times.' *Beverages.* 2018; 4(1): 15.

Clipper. 'Our sustainability naturally a better cup.' https://www.clipper-teas.com/our-story/unbleached-vs-bleached-bags/

Coca-Cola. 'Caffeine Counter.' http://www.tools.coca-cola.co.uk/gb/features/caffeine-counter

Cocoa Life. 'Cocoa Growing. The Challenge of Cocoa.' https://www.cocoalife.org/in-the-cocoa-origins/a-story-on-farming-cocoa-growing

Coffee and Health. 'Where coffee grows?' https://www.coffeeandhealth.org/all-about-coffee/where-coffee-grows/

Coffee and Health. 'Roasting and grinding.' https://www.coffeeandhealth.org/all-about-coffee/roasting-grinding/

Coffee and Health. 'Sports performance.' https://www.coffeeandhealth.org/topic-overview/sportsperformance/

Coffee chemistry. 'Differences between Arabica and Robusta Coffee.' https://www.coffeechemistry.com/general/agronomy/differences arabica-and-robusta-coffee 23 April 2015

Coffee Chemistry. 'Unlocking Coffee's Chemical Composition: Part 1.' https://www.coffeechemistry.com/library/coffee-science-publications/unlocking-coffee-s-chemical-composition-part-1

Coffee Confidential 'Decaffeination 101: Four ways to decaffeinate coffee.' https://coffeeconfidential.org/health/decaffeination/

Consumer Attorneys of California 'The McDonald's Hot Coffee Case.' https://www.caoc.org/?pg=facts

Cornell College of Agriculture and Life Sciences. 'Tannins: fascinating but sometimes dangerous molecules.' http://poisonousplants.ansci.cornell.edu/toxicagents/tannin.html

Counter Culture Coffee. 'Coffee Basics: How do you roast coffee?' https://counterculturecoffee.com/blog/coffee-basics-roasting 2 November 2017

Daily Mail. 'The danger of detox teas: Doctor warns most users have no idea the drinks can cause heart and bowel problems and even pregnancy.' https://www.dailymail.co.uk/health/article-3746884/The-danger-detox-teas-Doctor-warns-users-no-idea-drinks-cause-heart-bowel-problems-pregnancy.html 19 August 2016

Delimont, N. M., Haub, M. D. & Lindshield, B. L., 'The Impact of Tannin Consumption on Iron Bioavailability and Status: A Narrative Review.' *Current Developments in Nutrition.* 2017; 1(2): 1–12.

Dasanayake, A. P., Silverman, A. J. & Warnakulasuriya, S. 'Maté drinking and oral and oro-pharyngeal cancer: a systematic review and meta-analysis.' *Oral Oncol.* 2010; 46(2): 82–86.

Driftaway Coffee. 'What's the difference between Arabica and Robusta Coffee?' https://driftaway.coffee/arabica-robusta/

Driftaway Coffee. 'Why is coffee called a cup of joe?' https://driftaway.coffee/why-is-coffee-called-a-cup-of-joe/

Drug Bank. 'Theophylline.' https://www.drugbank.ca/drugs/DB00277

European Food Safety Agency. 'EFSA explains risk assessment Caffeine.' http://www.efsa.europa.eu/sites/default/files/corporate_publications/files/efsaexplainscaffeine150527.pdf

EFSA Panel on Dietetic Products, Nutrition and Allergies (NDA). 'Scientific Opinion on the substantiation of health claims related to Camelliasinensis (L.) Kuntze (tea), including catechins in green tea and tannins in black tea, and protection of DNA, proteins and lipids from oxidative damage (ID 1103, 1276, 1311, 1708, 2664), reduction of acid production in dental plaque (ID 1105, 1111), maintenance of bone (ID 1109), decreasing potentially pathogenic intestinal microorganisms (ID 1116), maintenance of vision (ID 1280), maintenance of normal blood pressure (ID 1546) and maintenance of normal blood cholesterol concentrations (ID 1113, 1114) pursuant to Article 13(1) of Regulation (EC) No 1924/2006.' *EFSA Journal.* 2010; 8(2): 1463.

EFSA. 'EFSA assesses safety of green tea catechins.' https://www.efsa.europa.eu/en/press/news/180418 18 April 2018

Eater. 'The McDonald's Hot Coffee Lawsuits Just Keep on Coming.' https://www.eater.com/2016/2/15/10996726/mcdonalds-hot-coffee-lawsuits-california-fresno 15 February 2016

Fitt, E., Pell, D. & Cole, D., 'Assessing caffeine intake in the United Kingdom diet.' *Food Chemistry.* 2013; 140(3): 421–426.

Food and Agriculture Organization of the United Nations. 'World tea production and trade Current and future development.' Rome 2015 http://www.fao.org/3/a-i4480e.pdf

Foodbev Media. 'HeatGenie raises $6m to bring its self-heating drink cans to market.' https://www.foodbev.com/news/heatgenie-raises-6m-bring-self-heating-drink-cans-market/ 11 June 2018

Food Component Database. '2-Ethylphenol (FDB005154).' http://foodb.ca/compounds/FDB005154

Franco, R., Oñatibia-Astibia, A. & Martínez-Pinilla, E., 'Health benefits of methylxanthines in cacao and chocolate.' *Nutrients.* 2013; 5(10): 4159–4173.

Frederick II of Prussia in 1777; quoted by Vallée, B. L., 'Alcohol in the Western World.' *Scientific American.* 1998; 278(6): 80–85.

Gambero, A. & Ribeiro, M. L., 'The positive effects of yerba maté (*Ilex para-guariensis*) in obesity.' *Nutrients.* 2015; 7(2): 730–750.

Gardner, E. J., Ruxton, C. H. S. & Leeds, A. R., 'Black tea – helpful or harmful? A review of the evidence.' *Eur J Clin Nutr.* 2007; 61: 3–18.

Go Ask Alice. 'Bagged tea versus loose leaf: Which is better?' http://goaska-lice.columbia.edu/answered-questions/bagged-tea-versus-loose-leaf-which-better

Green Tea Source. 'Where is Green Tea Consumed, and Produced the Most?' https://www.greenteasource.com/blog/where-green-tea-consumed-produced

Grosso, G., Godos, J., Galvano, F. & Giovannucci, E. L., 'Coffee, Caffeine, and Health Outcomes: An Umbrella Review.' *Annu Rev Nutr.* 2017; 37: 131–156.

Guayaki. 'Yerba Mate.' http://guayaki.com/mate/130/Yerba-Mate.html

Guo, Y., Zhi, F., Ping, C., Zhao, K., Xiang, H., Mao, Q., Wang, X. & Zhang, X., 'Green tea and the risk of prostate cancer: A systematic review and meta-analysis.' *Medicine.* 2017; 96(13): e6426.

Harvard Health Publishing. 'What is it about coffee?' https://www.health.harvard.edu/staying-healthy/what-is-it-about-coffee January 2012

Harvard Medical School. 'Helping Premature Babies Breathe Easier.' https://hms.harvard.edu/news/helping-premature-babies-breathe-easier 15 May 2014

Harvard School of Public Health. 'Carbohydrates and Blood Sugar.' https://www.hsph.harvard.edu/nutritionsource/carbohydrates/carbohydrates-and-blood-sugar/

Healthline. 'Does Hot Chocolate Have Caffeine? How It Compares to Other Beverages.' https://www.healthline.com/health/food-nutrition/does-hot-chocolate-have-caffeine#hot-chocolate-vs.-coffee

Higdon, J. V. & Frei, B., 'Tea catechins and polyphenols: health effects, metabolism, and antioxidant functions.' *Crit Rev Food Sci Nutr.* 2003; 43(1): 89–143.

Higgins, S., Straight, C. R. & Lewis, R. D., 'The Effects of Preexercise Caffeinated Coffee Ingestion on Endurance Performance: An Evidence-Based Review.' *Int J Sport Nutr Exerc Metab.* 2016; 26(3): 221–239.

Hodgson, J. M., Puddey, I. B., Woodman, R. J., et al. 'Effects of Black Tea on Blood Pressure: A Randomized Controlled Trial.' *Arch Intern Med.* 2012; 172(2): 186–188.

Horlicks. 'Our story.' http://www.horlicks.co.uk/story.html

Howstuffworks. 'How are coffee, tea and colas decaffeinated?' https://reci-pes.howstuffworks.com/question480.htm

iNews. 'Tetley follows PG Tips with pledge to eliminate all plastic from tea bags.'https://inews.co.uk/inews-lifestyle/food-and-drink/tetley-tea-follows-pg-tips-with-pledge-to-eliminate-all-plastic-from-tea-bags/ 21 March 2018

Initial. 'Tea Run.' https://www.initial.co.uk/washroom-news/2017/tea-run.html

Institute of Medicine (US) Standing Committee on the Scientific Evaluation of Dietary Reference Intakes. 'Dietary Reference Intakes for Calcium, Phosphorus, Magnesium, Vitamin D, and Fluoride.' Washington (DC): National Academies Press (US); 1997. 8, Fluoride.

International Cocoa Association. 'Processing cocoa.' https://www.icco.org/about-cocoa/processing-cocoa.html 7 June 2013

International Coffee Organization. http://www.ico.org

International Coffee Organization. 'Total production by all exporting coun-tries.' http://www.ico.org/prices/po-production.pdf

International Coffee Organization. 'Trade Statistics Tables.' http://www.ico.org/trade_statistics.asp?section=Statistics

Islami, F., Boffetta, P., Ren, J. S., Pedoeim, L., Khatib, D. & Kamangar, F., 'High-temperature beverages and foods and esophageal cancer risk – a systematic review.' Int J Cancer. 2009; 125(3): 491–524.

Ito En. 'Essential Green Varieties. How to Brew.' https://www.itoen.com/all-things-tea/major-varieties-tea

Johnson-Kozlow, M., Kritz-Silverstein, E. & Barrett-Connor, D. M., 'Coffee Consumption and Cognitive Function among Older Adults.' American Journal of Epidemiology. 2002; 156(9): 842–850.

Jurgens, T. & Whelan, A. M., 'Can green tea preparations help with weight loss?' Can Pharm J. 2014; 147(3): 159–160.

Kakumanu, N. & Sudhaker, D. R., 'Skeletal Fluorosis Due to Excessive Tea Drinking.' N Engl J Med. 2013; 368: 1140.

Keenan, E. K., Finnie, M. D. A., Jones. P. S., Rogers, P. J. & Priestley, C. M., 'How much theanine in a cup of tea? Effects of tea type and method of preparation.' Food Chemistry. 2011; 125(2): 588–594.

Khan, N. & Mukhtar, H., 'Tea and health: studies in humans.' Curr Pharm Des. 2013; 19(34): 6141–6147.

Kim, J. H., Desor, D., Kim, Y. T., Yoon, W. J., Kim, K. S., Jun, J. S., Pyun, K. H. & Shim, I., 'Efficacy of alphas1-casein hydrolysate on stress-related symptoms in women.' Eur J Clin Nutr. 2007; 61(4): 536–541.

Kim, Y. S., Kwak, S. M. & Myung, S. K., 'Caffeine intake from coffee or tea and cognitive disorders: a meta-analysis of observational studies.' *Neuroepidemiology*. 2015; 44(1): 51–63.

Know your phrase. 'Cup of Joe.' https://www.knowyourphrase.com/cup-of-joe

Kuura. 'Tea Dynamics: What Happens When We Steep Tea?' https://kuura.co/blogs/dispatch/tea-dynamics-i

Lipton Ice Tea. 'Products.' http://www.liptonicetea.com/en-GB/#products

Liu, Q. P., Wu, Y. F., Cheng, H. Y., Xia, T., Ding, H., Wang, H., Wang, Z. M. & Xu, Y., 'Habitual coffee consumption and risk of cognitive decline/dementia: A systematic review and meta-analysis of prospective cohort studies.' *Nutrition*. 2016; 32(6): 628–636.

Livertox Database, Drug Record. 'Green Tea Camellia Sinesis.' https://livertox.nih.gov/GreenTea.htm

Loria, D., Barrios, E. & Zanetti, R., 'Cancer and yerba mate consumption: a review of possible associations.' *Rev Panam Salud Publica*. 2009; 25(6): 530–539.

Lyngsø, J., Ramlau-Hansen, C. H., Bay, B., Ingerslev, H. J., Hulman, A. & Kesmodel, U. S., 'Association between coffee or caffeine consumption and fecundity and fertility: a systematic review and dose-response meta-analysis.' *Clin Epidemiol*. 2017; 9: 699–719.

McKay, D. L. & Blumberg, J. B., 'A review of the bioactivity and potential health benefits of chamomile tea (*Matricaria recutita L.*).' *Phytother Res*. 2006; 20(7): 519–530.

McKay, D. L. & Blumberg, J. B., 'A review of the bioactivity and potential health benefits of peppermint tea (*Mentha piperita L.*).' *Phytother Res*. 2006; 20(8): 619–633.

Madre Chocolate. 'Frequently asked questions.' http://madrechocolate.com/Frequently_Asked_Questions.html

Madzharov, A., Ye, N., Morrin, M. & Block, L., 'The impact of coffee-like scent on expectations and performance.' *Journal of Environmental Psychology*. 2018; 57: 83–86.

Make Chocolate Fair! 'Campaign Cocoa production in a nutshell.' https://makechocolatefair.org/issues/cocoa-production-nutshell

Mancini, E., Beglinger, C., Drewe, J., Zanchi, D., Lang, U. E. & Borgwardt, S., 'Green tea effects on cognition, mood and human brain function: A systematic review.' *Phytomedicine*. 2017; 34: 26–37.

Martin, M. A., Goya, L. & Ramos, S., 'Potential for preventive effects of cocoa and cocoa polyphenols in cancer.' *Food Chem Toxicol*. 2013; 56: 336–351.

Mayo Clinic. 'Caffeine content for coffee, tea, soda and more.' https://www.mayoclinic.org/healthy-lifestyle/nutrition-and-healthy-eating/in-depth/caffeine/art-20049372

Medicines and Healthcare products Regulatory Agency. 'Caffeine for apnoea of prematurity.' https://www.gov.uk/drug-safety-update/caffeine-for-apnoea-of-prematurity 11 December 2014

Mitchell, D. C., Knight, C. A., Hockenberry, J., Teplansky, R. & Hartman, T. J., 'Beverage caffeine intakes in the U.S.' *Food Chem Toxicol.* 2014; 63: 136–142.

Monteiro, J. P., Alves, M. G., Oliveira, P. F. & Silva, B. M., 'Structure-Bioactivity Relationships of Methylxanthines: Trying to Make Sense of All the Promises and the Drawbacks.' *Molecules.* 2016; 21(8): 974.

Muntons. 'What is Malt?' http://www.muntonsmalt.com/wp-content/uploads/2015/03/Health-benefits-of-Malt.pdf

National Center for Biotechnology Information. PubChem Database. 'Caffeine.' https://pubchem.ncbi.nlm.nih.gov/compound/2519

National Center for Biotechnology Information. PubChem Database. 'Ethyl acetate.' https://pubchem.ncbi.nlm.nih.gov/compound/8857

National Center for Biotechnology Information. PubChem Database. 'Methylene chloride.' https://pubchem.ncbi.nlm.nih.gov/compound/6344

National Center for Biotechnology Information. PubChem Database. 'Theophylline.' https://pubchem.ncbi.nlm.nih.gov/compound/2153

National Center for Biotechnology Information. PubChem Database. 'Trigonelline' https://pubchem.ncbi.nlm.nih.gov/compound/5570

National Center for Biotechnology Information. PubChem Database. 'Serotonin.' https://pubchem.ncbi.nlm.nih.gov/compound/5202

National Center for Biotechnology Information. PubChem Database. 'Tryptophan.' https://pubchem.ncbi.nlm.nih.gov/compound/6305

National Coffee Association USA. 'What is coffee?' http://www.ncausa.org/About-Coffee/What-is-Coffee

National Coffee Association USA. 'How to brew coffee.' https://www.ncausa.org/About-Coffee/How-to-Brew-Coffee

National Coffee Association USA. 'Coffee roast guide.' http://www.ncausa.org/About-Coffee/Coffee-Roasts-Guide

National Osteoporosis Foundation. 'Frequently Asked Questions.' https://www.nof.org/patients/patient-support/faq/

Nelson, M. & Poulter, J., 'Impact of tea drinking on iron status in the UK: a review.' *J Hum Nutr Diet.* 2004; 17(1): 43–54.

Nestle. 'Meet the Milo supermen who inspired our super brand.' https://www.nestle.com/aboutus/history/nestle-company-history/milo

Newsweek. 'Coffee brain boost: smell alone can bring higher math test scores.' https://www.newsweek.com/coffee-brain-boost-smell-alone-can-bring-higher-math-test-scores-researchers-1029044

NHS 'Should I limit caffeine during pregnancy?' https://www.nhs.uk/common-health-questions/pregnancy/should-i-limit-caffeine-during-pregnancy/ 2 May 2018

Nieber, K., 'The Impact of Coffee on Health.' *Planta Med.* 2017; 83(16): 1256–1263.

Norfolk Dental Specialists. 'Causes, Prevention and Treatment of Tooth Staining.' https://www.ndspecialists.uk/news/causes-prevention-and-treatment-of-tooth-staining

North Star Coffee Roasters. 'Roasting Coffee: Light, Medium and Dark Roasts Explained.' https://www.northstarroast.com/roasting-coffee-light-medium-dark/

Panza, F., Solfrizzi, V., Barulli, M. R., Bonfiglio, C., Guerra, V., Osella, A., Seripa, D., Sabbà, C., Pilotto, A. & Logroscino, G., 'Coffee, tea, and caffeine consumption and prevention of late-life cognitive decline and dementia: a systematic review.' *J Nutr Health Aging.* 2015; 19(3): 313–328.

Peng, C. Y., Zhu, X. H., Hou, R. Y., Ge, G. F., Hua, R. M., Wan, X. C. & Cai, H. M., 'Aluminum and Heavy Metal Accumulation in Tea Leaves: An Interplay of Environmental and Plant Factors and an Assessment of Exposure Risks to Consumers.' *J Food Sci.* 2018; 83(4): 1165–1172.

Phongnarisorn, B., Orfila, C., Holmes, M. & Marshall, L. J., 'Enrichment of Biscuits with Matcha Green Tea Powder: Its Impact on Consumer Acceptability and Acute Metabolic Response.' *Foods.* 2018; 7(2): 17.

Poole, R., Kennedy, O. J., Roderick, P., Fallowfield, J. A., Hayes, P. C. & Parkes, J., 'Coffee consumption and health: umbrella review of meta-analyses of multiple health outcomes.' *BMJ.* 2017; 359: j5024.

Pucciarelli, D. L., 'Cocoa and heart health: a historical review of the science.' *Nutrients.* 2013; 5(10): 3854–3870.

Rainforest Alliance. 'Chocolate: The Journey From Beans to Bar.' https://www.rainforest-alliance.org/pictures/chocolate-from-bean-to-bar

Red Bull. 'Caffeine.' http://energydrink-uk.redbull.com/red-bull-caffeine-content

Reygaert, W. C., 'An Update on the Health Benefits of Green Tea.' *Beverages* 2017; 3(1): 6.

Richards, G., Smith, A. P., 'Caffeine Consumption and General Health in Secondary School Children: A Cross-sectional and Longitudinal Analysis.' *Front Nutr.* 2016; 3: 52.

Rodríguez-Artalejo, F. & López-García, E., 'Coffee Consumption and Cardiovascular Disease: A Condensed Review of Epidemiological Evidence and Mechanisms.' *J Agric Food Chem.* 2018; 66(21): 5257–5263.

Rossi, T., Gallo, C., Bassani, B., Canali, S., Albini, A. & Bruno, A., 'Drink your prevention: beverages with cancer preventive phytochemicals.' *Pol Arch Med Wewn.* 2014; 124(12): 713–722.

Royal Botanic Gardens Kew. '*Camellia sinensis.*' http://powo.science.kew.org/taxon/urn:lsid:ipni.org:names:828548-1

Royal Society of Chemistry. 'How to make a Perfect Cup of Tea.' http://www.academiaobscura.com/wp-content/uploads/2014/10/RSC-tea-guidelines.pdf

Sainsbury's. 'Ovaltine Malted Drink, Original 300g.' https://www.sainsburys.co.uk/shop/gb/groceries/ovaltine-malted-drink--original-300g

Schwalfenberg, G., Genuis, S. J. & Rodushkin, I., 'The benefits and risks of consuming brewed tea: beware of toxic element contamination.' *J Toxicol.* 2013: 370460.

Schubert, M. M., Irwin, C., Seay, R. F., Clarke, H. E., Allegro, D. & Desbrow, B., 'Caffeine, coffee, and appetite control: a review.' *Int J Food Sci Nutr.* 2017; 68(8): 901–912.

Schulze, J., Melzer, L., Smith, L. & Teschke, R., 'Green Tea and Its Extracts in Cancer Prevention and Treatment.' *Beverages.* 2017; 3(1): 17.

Scientific American. 'How is caffeine removed to produce decaffeinated coffee?' https://www.scientificamerican.com/article/how-is-caffeine-removed-t/

Skinner, T. L., Jenkins, D. G., Taaffe, D. R., Leveritt, M. D. & Coombes, J. S., 'Coinciding exercise with peak serum caffeine does not improve cycling performance.' *J Sci Med Sport.* 2013; 16(1): 54–59.

Smithsonian.com 'A Hot Drink on a Hot Day Can Cool You Down.' https://www.smithsonianmag.com/science-nature/a-hot-drink-on-a-hot-day-can-cool-you-down-1338875/

Snopes. 'Why is Coffee Called a "Cup of Joe"?' https://www.snopes.com/fact-check/cup-of-joe/ 9 January 2009

Solfrizzi, V., Panza, F., Imbimbo, B. P., D'Introno, A., Galluzzo, L., Gandin, C., Misciagna, G., Guerra, V., Osella, A., Baldereschi, M., Di Carlo, A., Inzitari, D., Seripa, D., Pilotto, A., Sabbá. C., Logroscino, G., Scafato, E.; Italian Longitudinal Study on Aging Working Group. 'Coffee Consumption Habits and the Risk of Mild Cognitive Impairment: The Italian Longitudinal Study on Aging.' *J Alzheimers Dis.* 2015; 47(4): 889–899.

Song, J., Xu, H., Liu, F. & Feng, L., 'Tea and cognitive health in late life: current evidence and future directions.' *J Nutr Health Aging.* 2012; 16(1): 31–34.

Stadheim, H. K., Spencer, M., Olsen, R. & Jensen, J., 'Caffeine and performance over consecutive days of simulated competition.' *Med Sci Sports Exerc.* 2014; 46(9): 1787–1796.

Standley, L., Winterton, P., Marnewick, J. L., Gelderblom, W. C., Joubert, E. &

Britz, T. J., 'Influence of processing stages on antimutagenic and antioxidant potentials of rooibos tea.' *J Agric Food Chem.* 2001; 49(1): 114–117.

Starbucks. 'Summer 2 2018 Beverage Ingredients.' https://globalassets.starbucks.com/assets/68FC43D2BE3244C9A70EE30EA57B4880.pdf

Statista 'The British drink less tea but more coffee.' https://www.statista.com/chart/10196/coffee-and-tea-purchases-in-the-uk/ 10 July 2017

Statista. 'Annual per capita tea consumption worldwide as of 2016.' https://www.statista.com/statistics/507950/global-per-capita-tea-consumption-by-country/

Statista. 'Global beverage sales share from 2011 to 2016, by beverage type.' https://www.statista.com/statistics/232773/forecast-for-global-beverage-sales-by-beverage-type/

St-Onge, M. P., Mikic, A. & Pietrolungo, C. E., 'Effects of Diet on Sleep Quality.' *Adv Nutr.* 2016; 7(5): 938–949.

Supermarketnews. 'Self-heating coffee in a can.' http://supermarketnews.co.nz/self-heating-coffee-in-a-can/ 28 August 2018

Suzuki, Y., Miyoshi, N. & Isemura, M., 'Health-promoting effects of green tea.' *Proc Jpn Acad Ser B Phys Biol Sci.* 2012; 88(3): 88–101.

Tea Advisory Panel. 'Health and Wellbeing.' https://www.teaadvisorypanel.com/tea/health-wellbeing

Tea Association of the USA. 'Tea fact sheet – 2018–2019.' http://www.teausa.com/14655/tea-fact-sheet

Teatulia. 'What is Chai?' https://www.teatulia.com/tea-varieties/what-is-chai.htm

Tea Class. 'Yerba Mate.' https://www.teaclass.com/lesson_0309.html

Tea Class. 'Types of Tea.' https://www.teaclass.com/lesson_0102.html

Tea Metabolome Database. http://pcsb.ahau.edu.cn:8080/TCDB/f

Teatulia. 'Tea Processing.' https://www.teatulia.com/tea-101/tea-processing.htm

Temple, J. L., Bernard, C., Lipshultz, S. E., Czachor, J. D., Westphal, J. A. & Mestre, M. A., 'The Safety of Ingested Caffeine: A Comprehensive Review.' *Frontiers in Psychiatry.* 2017; 8. https://doi.org/10.3389/fpsyt.2017.00080

Tesco. 'Ovaltine Original Add Milk Drink 300G.' https://www.tesco.com/groceries/en-GB/products/258492318

Tesco. 'Nestle Milo 400G.' https://www.tesco.com/groceries/en-GB/products/259536574

The Atlantic. 'Map: The Countries That Drink the Most Tea. Move over, China. Turkey is the real titan of tea.' https://www.theatlantic.com/international/archive/2014/01/map-the-countries-that-drink-the-most-tea/283231/ 21 January 2014

The Conversation. 'What science says about getting the most out of your tea.' http://theconversation.com/what-science-says-about-getting-the-most-out-of-your-tea-75767 18 April 2017

The Conversation. 'Self-heating drinks cans return – here's how they work.' http://theconversation.com/self-heating-drinks-cans-return-heres-how-they-work-98476 20 June 2018

The Co-operative Group. 'The New "Green" Tea: Co-op Brews Up Solution To Plastic Tea Bags'. https://www.co-operative.coop/media/news-releases/the-new-green-tea-co-op-brews-up-solution-to-plastic-tea-bags 30 January 2018

The Huffington Post. 'McDonald's Hot Coffee Controversy Is Back With Another Burn Lawsuit.' https://www.huffingtonpost.co.uk/entry/mcdonalds-hot-coffee-suit_n_4192626 12 June 2017

The Huffington Post. 'The McDonalds' Coffee Case.' https://www.huffingtonpost.com/darryl-s-weiman-md-jd/the-mcdonalds-coffee-case_b_14002362.html 7 January 2018

The Pherobase. '2-ethylphenol.' http://www.pherobase.com/database/compound/compounds-detail-2-ethylphenol.php?isvalid=yes

TuftsNow. 'Does tea lose its health benefits if it's been stored a long time? And is it better to use loose tea or tea bags?' http://now.tufts.edu/articles/tea-health-benefits-storage-time 26 May 2011

Twinings. 'Chinese Jasmine Green.' https://www.twinings.co.uk/tea/loose-tea/chinese-jasmine-green

Twinings. 'How is Tea Made.' https://www.twinings.co.uk/about-tea/how-is-tea-made

UK Tea and Infusions Association. 'The History of the Tea Bag.' https://www.tea.co.uk/the-history-of-the-tea-bag

UK Tea and Infusions Association. 'Tea Facts.' https://www.tea.co.uk/tea-facts

Unachukwu, U. J., Ahmed, S., Kavalier, A., Lyles, J. T. & Kennelly, E. J., 'White and green teas (*Camellia sinensis* var. *sinensis*): variation in phenolic, methylxanthine, and antioxidant profiles.' *J Food Sci.* 2010; 75(6): C541–548.

University of Cambridge. 'Bovril – a very beefy (and British) love affair.' https://www.cam.ac.uk/research/news/bovril-a-very-beefy-and-british-love-affair 5 July 2013

Verster, J. C. & Koenig, J., 'Caffeine intake and its sources: A review of national representative studies.' *Critical Reviews in food Science and Nutrition.* 2018; 58(8): 1250–1259.

Vuong, Q. V., 'Epidemiological Evidence Linking Tea Consumption to Human Health: A Review.' *Critical Reviews in Food Science and Nutrition.* 2014; 54(4): 523–536.

Vuong, Q. V., Tan, S. P., Stathopoulos, C. E. & Roach, P. D., 'Improved extraction of green tea components from teabags using the microwave oven.' *Journal of Food Composition and Analysis.* 2012; 27(1): 95–101.

Wang, D., Chen, C., Wang, Y., Liu, J. & Lin, R., 'Effect of Black Tea Consumption on Blood Cholesterol: A Meta-Analysis of 15 Randomized Controlled Trials.' *PLoS ONE.* 2014; 9(9): e107711.

Waugh, D. T., Potter, W., Limeback, H. & Godfrey, M., 'Risk Assessment of Fluoride Intake from Tea in the Republic of Ireland and its Implications for Public Health and Water Fluoridation.' *Int J Environ Res Public Health.* 2016; 13(3): 259.

WebMD 'Decaf coffee isn't caffeine free.' https://www.webmd.com/diet/news/20061011/decaf-coffee-isnt-caffeine-free 11 October 2006

Whayne, T. F., Jr., 'Coffee: A Selected Overview of Beneficial or Harmful Effects on the Cardiovascular System?' *Curr Vasc Pharmacol.* 2015; 13(5): 637–648.

Wierzejska, R., 'Tea and health – a review of the current state of knowledge.' *Przegl Epidemiol.* 2014; 68(3): 501–506, 595–599.

Williams, J., Kellett, J., Roach, P. D., McKune, A., Mellor, D., Thomas, J. & Naumovski, N., 'Review l-Theanine as a Functional Food Additive: Its Role in Disease Prevention and Health Promotion.' *Beverages.* 2016; 2(2): 13.

Willems, M. E. T., Şahin, M. A. & Cook, M. D., 'Matcha Green Tea Drinks Enhance Fat Oxidation During Brisk Walking in Females.' *Int J Sport Nutr Exerc Metab.* 2018; 28(5): 536–541.

Winston, A., Hardwick, E. & Jaberi, N., 'Neuropsychiatric effects of caffeine.' *Advances in Psychiatric Treatment.* 2005; 11(6): 432–439.

Women's Health. 'If You've Been Tempted By The Quick Wins of Detox Tea, Read This.' https://www.womenshealthmag.com/uk/food/healthy-eating/a707711/detox-tea/ 28 November 2018

World Cocoa Foundation. https://www.worldcocoafoundation.org

World Cocoa Foundation. 'Cocoa glossary.' https://www.worldcocoafoundation.org/cocoa-glossary/

World Cocoa Foundation. 'History of Cocoa.' http://www.worldcocoafoundation.org/about-cocoa/history-of-cocoa/ 15 August 2018

Wu, L., Sun, D. & He, Y., 'Coffee intake and the incident risk of cognitive disorders: A dose-response meta-analysis of nine prospective cohort studies.' *Clin Nutr.* 2017; 36(3): 730–736.

Yang, J., Mao, Q. X., Xu, H. X., Ma, X. & Zeng, C. Y., 'Tea consumption and risk of type 2 diabetes mellitus: a systematic review and meta-analysis update.' *BMJ Open.* 2014; 4(7): e005632.

Yuan, J. M., 'Cancer prevention by green tea: evidence from epidemiologic studies.' *Am J Clin Nutr.* 2013; 98(6 Suppl): 1676S–1681S.

Yue, Y., Chu, G. X., Liu, X. S., Tang, X., Wang, W., Liu, G. J., Yang, T., Ling, T. J., Wang, X. G., Zhang, Z. Z., Xia, T., Wan, X. C. & Bao, G. H., 'TMDB: a literature-curated database for small molecular compounds found from tea.' *BMC Plant Biol.* 2014; 16(14): 243.

Zhao, Y., Asimi, S., Wu, K., Zheng, J. & Li, D., 'Black tea consumption and serum cholesterol concentration: Systematic review and meta-analysis of randomized controlled trials.' *Clin Nutr.* 2015; 34(4): 612–619.

Zhang, Y. F., Xu, Q., Lu, J., Wang, P., Zhang, H. W., Zhou, L., Ma, X. Q. & Zhou, Y. H., 'Tea consumption and the incidence of cancer: a systematic review and meta-analysis of prospective observational studies.' *Eur J Cancer Prev.* 2015; 24(4): 353–362.

Zhou, A., Taylor, A. E., Karhunen, V., et al. 'Habitual coffee consumption and cognitive function: a Mendelian randomization meta-analysis in up to 415,530 participants.' *Sci Rep.* 2018; 8(1): 7526.

4. COLD DRINKS (NON-ALCOHOLIC)

AG Barr. 'The phenomenal A.G. Barr story.' https://www.agbarr.co.uk/about-us/our-history/timeline/

Ahlawat, K. S. & Khatkar, B. S., 'Processing, food applications and safety of aloe vera products: a review.' *J Food Sci Technol.* 2011; 48(5): 525–533.

Air Pollution Information System. 'Sulphur Dioxide.' http://www.apis.ac.uk/overview/pollutants/overview_SO2.htm

Al-Shaar, L., Vercammen, K., Lu, C., Richardson, S. et al. 'Health Effects and Public Health Concerns of Energy Drink Consumption in the United States: A Mini-Review.' *Frontiers in Public Health.* 2017; 5: 225.

American Society for Nutrition. 'Are all sugars created equal? Let's talk fructose metabolism.' https://nutrition.org/sugars-created-equal-lets-talk-fructose-metabolism/ 3 December 2015

Annie Andre. '7 strange table manners around the world: burping, farting+.' https://www.annieandre.com/world-table-manners-etiquette/

ASCIA. 'Sulfite sensitivity.' https://www.allergy.org.au/patients/product-allergy/sulfite-allergy 2014

Asgari-Taee, F., Zerafati-Shoae, N. & Dehghani, M. et al. 'Association of sugar sweetened beverages consumption with non-alcoholic fatty liver disease: a systematic review and meta-analysis.' *Eur J Nutr.* May 2018: 1–11.

Australian Government Department of the Environment and Energy. 'What is sulfur dioxide?' http://www.environment.gov.au/protection/publications/factsheet-sulfur-dioxide-so2 2005

BBC. 'Coca-Cola "in talks" over cannabis-infused drinks.' https://www.bbc.co.uk/news/business-45545233 17 September 2018

BBC. 'High sport drink use among young teens "risk to health".' https://www.bbc.co.uk/news/uk-wales-south-east-wales-36638596 27 June 2016

BDA. 'Consumption of energy drinks by young people – what is the evidence?' https://www.bda.uk.com/dt/articles/energy_drinks_young_people

Beezhold, B. L., Johnston, C. S. & Nochta, K. A., 'Sodium Benzoate-Rich Beverage Consumption is Associated with Increased Reporting of ADHD Symptoms in College Students: A Pilot Investigation.' *Journal of Attention Disorders.* 2014; 18(3): 236–241.

Berkeley Wellness. 'Is high fructose corn syrup worse than regular sugar?' http://www.berkeleywellness.com/healthy-eating/nutrition/article/high-fructose-corn-syrup-worse-regular-sugar 7 June 2017

Beverage Daily. 'Cactus, birch, lychee and lemongrass: Soft drink consumers turn to natural flavors and functional innovations.' https://www.beveragedaily.com/Article/2016/01/06/Soft-drinks-turn-to-natural-flavors-and-functional-innovations# 6 January 2016

Bezkorovainy, A., 'Probiotics: determinants of survival and growth in the gut.' *Am J Clin Nutr.* 2001; 73(2): 399s–405s.

Bian, X., Chi, L., Gao, B., Tu, P., Ru, H. & Lu, K., 'The artificial sweetener acesulfame potassium affects the gut microbiome and body weight gain in CD-1 mice.' *PLoS One.* 2017;12(6): e0178426.

Bleakley, S. & Hayes, M., 'Algal Proteins: Extraction, Application, and Challenges Concerning Production.' *Foods.* 2017; 6(5): 33.

BMJ. 'Research news. Fructose may be making us eat more.' https://www.bmj.com/content/346/bmj.f74 9 January 2013

Brink-Elfegoun, T., Ratel, S., Leprêtre, P. M., et al. 'Effects of sports drinks on the maintenance of physical performance during 3 tennis matches: a randomized controlled study.' *J Int Soc Sports Nutr.* 2014; 11: 46.

British Nutrition Foundation. 'Liquids.' https://www.nutrition.org.uk/nutritionscience/nutrients-food-and-ingredients/liquids.html?limit=1&start=3 July 2009

British Nutrition Foundation. 'Minerals and trace elements.' https://www.nutrition.org.uk/nutritionscience/nutrients-food-and-ingredients/minerals-and-trace-elements.html?limit=1&start=14

British Nutrition Foundation. 'Nutrition requirements.' https://www.nutrition.org.uk/attachments/article/234/Nutrition%20Requirements_Revised%20Oct%202016.pdf October 2016

British Nutrition Foundation. 'Should children be drinking energy drinks?' https://www.nutrition.org.uk/nutritioninthenews/headlines/children-energydrinks.html 18 January 2018

British Soft Drinks Association. 'Carbonated drinks.' http://www.british-softdrinks.com/Carbonated-Fizzy-Drinks

British Soft Drinks Association. 'Dilutables.' http://www.britishsoftdrinks.com/Dilutables

British Soft Drinks Association. 'Fruit juices.' http://www.britishsoftdrinks.com/Fruit-Juices

British Soft Drinks Association. 'Ingredients.' http://www.britishsoftdrinks.com/Ingredients

British Soft Drinks Association. 'Soft drinks.' http://www.britishsoftdrinks.com/soft-drinks

Broad, E. M. & Rye, L. A., 'Do current sports nutrition guidelines conflict with good oral health?' *Gen Dent.* 2015; 63(6): 18–23.

Brown, C. J., Smith, G., Shaw, L., Parry, J. & Smith, A. J., 'The erosive potential of flavoured sparkling water drinks.' *International Journal of Paediatric Dentistry.* 2007; 17: 86–91.

Calvo, M. S. & Tucker, K. L., 'Is phosphorus intake that exceeds dietary requirements a risk factor in bone health?' *Ann. N.Y. Acad. Sci.* 2013; 1301: 29–35.

Cannabis Drinks Expo. 'Home.' http://cannabisdrinksexpo.com

CNN Health. 'What are natural flavors, really?' https://edition.cnn.com/2015/01/14/health/feat-natural-flavors-explained/index.html 14 January 2015

Coca-Cola. 'Are there any additives in Coca-Cola?' https://www.coca-cola.co.uk/faq/ingredients/do-you-use-additives-or-preservatives-in-coca-cola

Coca-Cola. 'Brands: Sprite.' https://www.coca-cola.co.uk/drinks/sprite/

Coca-Cola. 'Coca-Cola Original Taste.' https://www.coca-cola.co.uk/drinks/coca-cola/coca-cola

Coca-Cola. 'How many calories are there in a 330ml can of Coca-Cola original taste?' https://www.coca-cola.co.uk/faq/calories-in-330ml-can-of-coca-cola

Coca-Cola. 'Let's talk about the government's soft drinks tax and what that means for our drinks.' https://www.coca-cola.co.uk/blog/lets-talk-about-soft-drinks-tax? 5 April 2018

Coca-Cola. 'What are the ingredients of a Coca-Cola Classic?' https://www.coca-cola.co.uk/faq/ingredients/what-are-the-ingredients-of-coca-cola-classic

Coconut Knowledge Centre Singapore. http://www.lankacoconutgrowers.com/pdf/Coconut_Knowledge_Center.pdf

Cohen, D., 'The truth about sports drinks.' *BMJ.* 2012; 345: e4737.

Daily Beast. 'The scoop on sprirulina: should you eat this microalgae?' https://www.thedailybeast.com/the-scoop-on-spirulina-should-you-eat-this-microalgae 20 March 2015

Davies, R., 'Effect of fructose on overeating visualised.' *The Lancet Diabetes & Endocrinology.* 2013; 1: S7.

Department of Health. 'Nutrient analysis of fruit and vegetables. Summary report.' March 2013.

Derlet, R. W. & Albertson, T. E., 'Activated charcoal – past, present and future.' *The Western Journal of Medicine.* 1986; 145(4): 493–496.

Di, R., Huang, M. T. & Ho, C. T., 'Anti-inflammatory Activities of Mogrosides from Momordica grosvenori in Murine Macrophages and a Murine Ear Edema Model.' *J. Agric. Food Chem.* 2011; 59(13): 7474–7481.

Drugs and Lactation Database (LactMed). Bethesda (MD): National Library of Medicine (US); 2006–. *Turmeric.* [Updated 2019 Jan 7]. Available from: https://www.ncbi.nlm.nih.gov/books/NBK501846/

Du, M., Tugendhaft, A., Erzse, A. & Hofman, K. J., 'Sugar-Sweetened Beverage Taxes: Industry Response and Tactics.' *Yale J Biol Med.* 2018; 91(2): 185–190.

Eccles, R., Du-Plessis, L., Dommels, Y. & Wilkinson, J. E., 'Cold pleasure. Why we like ice drinks, ice-lollies and ice cream.' *Appetite.* 2013; 71: 357–336.

EFFA. 'EFFA Guidance Document on the EC Regulation on Flavourings.' http://www.effa.eu/docs/default-source/guidance-documents/effa_guidance-document-on-the-ec-regulation-on-flavourings.pdf?sfvrsn=2

EFSA. 'EFSA assesses new aspartame study and reconfirms its safety.' https://www.efsa.europa.eu/en/press/news/060504 4 May 2006

EFSA Panel on Food Additives and Nutrient Sources added to Food (ANS). 'Scientific Opinion on the re-evaluation of aspartame (E 951) as a food additive.' *EFSA Journal.* 2013; 11(12): 3496.

EFSA Panel on Food Additives and Nutrient Sources added to Food (ANS). 'Scientific opinion on the re-evaluation of dimethyl dicarbonate (DMDC, E 242) as a food additive.' *EFSA Journal.* 2015; 13(12): 4319.

EFSA Panel on Food Additives and Nutrient Sources added to Food (ANS). 'Scientific Opinion on the re-evaluation of sulfur dioxide (E 220), sodium sulfite (E 221), sodium bisulfite (E 222), sodium metabisulfite (E 223), potassium metabisulfite (E 224), calcium sulfite (E 226), calcium bisulfite (E 227) and potassium bisulfite (E 228) as food additives.' *EFSA Journal.* 2016; 14(4): 4438.

EFSA Panel on Food Additives and Nutrient Sources added to Food (ANS). 'Scientific Opinion on the safety of steviol glycosides for the proposed uses as a food additive.' *EFSA Journal.* 2010; 8(4): 1537.

Emmins, C., *Soft Drinks. Their origins and history*. Shire Album 269. Shire Publications Ltd.

Ernst, E., 'Kombucha: a systematic review of the clinical evidence.' *Forsch Komplementarmed Klass Naturheilkd*. 2003; 10(2): 85–87.

Eufic. 'Acidity regulators: The multi-task players.' https://www.eufic.org/en /whats-in-food/article/acidity-regulators-the-multi-task-players 1 December 2004

European Commission Scientific Committee on Food. 'Opinion of the Scientific Committee on Food on sucralose.' SCF/CS/ADDS/EDUL/ 190 Final. 12 September 2000

European Commission Scientific Committee on Food. 'Opinion on saccharin and its sodium, potassium and calcium salts.' CS/ADD/EDUL/148- FINAL. February 1997

FDA. 'Carbonated soft drinks: what you should know.' https://www.fda.gov /food/ingredientspackaginglabeling/foodadditivesingredients/ ucm232528.htm 3 January 2018

FDA. 'High fructose corn syrup questions and answers.' https://www.fda. gov/food/ingredientspackaginglabeling/foodadditivesingredients/ ucm324856.htm 4 April 2018

Field, A. E., Sonneville, K. R., Falbe, J., et al. 'Association of sports drinks with weight gain among adolescents and young adults.' *Obesity* (Silver Spring). 2014; 22(10): 2238–2243.

Flood-Obbagy, J. E. & Rolls, B. J., 'The effect of fruit in different forms on energy intake and satiety at a meal.' *Appetite*. 2008; 52(2): 416–422.

Food Ingredients Online. 'Beverage stabilizers.' https://www.foodingredi- entsonline.com/doc/beverage-stabilizers-0001I5 November 2000

Food Standards Agency. 'Food additives.' https://www.food.gov.uk/safety- hygiene/food-additives 9 January 2018

Forbes. '5 more locations pass soda taxes: what's next for big soda?' https:/ /www.forbes.com/sites/brucelee/2016/11/14/5-more-locations-pass- soda-taxes-whats-next-for-big-soda/#1b86d0ded192 14 November 2016

Ford, A. C., Harris, L. A., Lacy, B. E., Quigley, E. M. M. & Moayyedi, P., 'Systematic review with meta-analysis: the efficacy of prebiotics, probiotics, synbiotics and antibiotics in irritable bowel syndrome.' *Aliment Pharmacol Ther*. 2018; 48: 1044–1060.

Gedela, M., Potu, K. C., Gali, V. L., Alyamany, K. & Jha, L. K., 'A Case of Hepatotoxicity Related to Kombucha Tea Consumption.' *S D Med*. 2016; 69(1): 26–28.

Gov. UK. 'Soft Drinks Industry Levy comes into effect.' https://www.gov.uk /government/news/soft-drinks-industry-levy-comes-into-effect 5 April 2018

Greenwalt, C. J., Steinkraus, K. H. & Ledford, R. A., 'Kombucha, the fermented tea: microbiology, composition, and claimed health effects.' *J Food Prot.* 2000; 63(7): 976–981.

Gupta, S. C., Patchva, S. & Aggarwal, B. B., 'Therapeutic roles of curcumin: lessons learned from clinical trials.' *AAPS J.* 2012; 15(1): 195–218.

Hamman, J. H., 'Composition and Applications of *Aloe vera* Leaf Gel.' *Molecules.* 2008; 13(8): 1599–1616.

Healthline. 'Longan fruit vs. lychee: health benefits, nutrition information and uses.' https://www.healthline.com/health/longan-fruit-vs-lychee-benefits#takeaway 22 June 2017

Hewlings, S. J. & Kalman, D. S., 'Curcumin: A Review of Its' Effects on Human Health.' *Foods.* 2017; 6(10): 92.

History of Soft Drinks 'How are soft drinks made?' http://www.historyof-softdrinks.com/making-soda/how-soft-drinks-are-made/

Holbourn, A. & Hurdman, J., 'Kombucha: is a cup of tea good for you?' *BMJ Case Rep.* 2 December 2017; 2017: pii: bcr-2017-221702.

Holland & Barrett. 'Invo Pure Coconut Water.' https://www.hollandandbarrett.com/shop/product/invo-pure-coconut-water-60028523

Hu, F. B., 'Resolved: there is sufficient scientific evidence that decreasing sugar-sweetened beverage consumption will reduce the prevalence of obesity and obesity-related diseases.' *Obes Rev.* 2013; 14(8): 606–619.

Hu, F. B. & Malik, V. S., 'Sugar-sweetened beverages and risk of obesity and type 2 diabetes: epidemiologic evidence.' *Physiol Behav.* 2010; 100(1): 47–54.

Huang, C., Huang, J., Tian, Y. et al. 'Sugar sweetened beverages consumption and risk of coronary heart disease: A meta-analysis of prospective studies.' *Atherosclerosis.* 2014; 234(1): 11–16.

Huffington Post. 'Here's why "maple water" isn't the new anything.' https://www.huffingtonpost.co.uk/entry/maple-water_n_5606092 24 July 2014

HYET Sweet. 'Sweetener system for soft drinks.' https://www.hyetsweet.com/wp-content/themes/HyetSweet/includes/img/Leaflet_SSfSD_HYET_Sweet.pdf

IARC. 'Saccharin and its salts.' IARC Monographs Volume 73: 517–624.

Imamura, F., O'Connor, L., Ye, Z., et al. 'Consumption of sugar sweetened beverages, artificially sweetened beverages, and fruit juice and incidence of type 2 diabetes: systematic review, meta-analysis, and estimation of population attributable fraction.' *BMJ.* 2015; 351: h3576.

The Independent. 'Food agency calls for ban on six artificial colours.' https://www.independent.co.uk/life-style/food-and-drink/news/food-agency-calls-for-ban-on-six-artificial-colours-807806.html 11 April 2008

The Independent. 'Irn-Bru: 15 things you didn't know about Scotland's national drink.' https://www.independent.co.uk/life-style/food-and-drink/irn-bru-things-what-is-didnt-know-recipe-change-ag-barr-scotland-favourite-soft-drink-can-a8143301.html 5 January 2018

The Independent. 'The real thing? Historian publishes Coca-Cola's 'secret formula'.' https://www.independent.co.uk/news/world/americas/the-real-thing-historian-publishes-coca-colas-secret-formula-8619076.html 16 May 2017

Institute of Medicine (US) Panel on Micronutrients. *Dietary Reference Intakes for Vitamin A, Vitamin K, Arsenic, Boron, Chromium, Copper, Iodine, Iron, Manganese, Molybdenum, Nickel, Silicon, Vanadium, and Zinc*. Washington (DC): National Academies Press (US). 2001; 10, Manganese. Available from: https://www.ncbi.nlm.nih.gov/books/NBK222332/

International Food Information Council Foundation. 'Everything you need to know about sucralose.' http://www.foodinsight.org/articles/everything-you-need-know-about-sucralose 26 November 2018

Jamwal, R., 'Bioavailable curcumin formulations: A review of pharmacokinetic studies in healthy volunteers.' *J Integr Med*. 2018; 16(6): 367–374.

Johnson, L. A., Foster, D. & McDowell, J. C., 'Energy Drinks: Review of Performance Benefits, Health Concerns, and Use by Military Personnel.' *Military Medicine*. 2014; 179(4): 375–380.

Kalman, D. S., Feldman, S., Krieger, D. R. & Bloomer, R. J., 'Comparison of coconut water and a carbohydrate-electrolyte sport drink on measures of hydration and physical performance in exercise-trained men.' *J Int Soc Sports Nutr*. 2012; 9(1): 1.

Kleerebezem, M., Binda, S. & Bron, P. A., 'Understanding mode of action can drive the translational pipeline towards more reliable health benefits for probiotics.' *Curr Opin Biotechnol*. 2015; 56: 55–60.

Kole, A. S., Jones, H. D. & Christensen, R., 'A Case of Kombucha Tea Toxicity.' *Journal of Intensive Care Medicine*. 2009; 24(3): 205–207.

Korea.net. 'Korean recipes: Traditional drinks keep you healthy in winter.' http://www.korea.net/NewsFocus/Culture/view?articleId=131900 15 January 2016

Kregiel, D., 'Health safety of soft drinks: contents, containers, and microorganisms.' *Biomed Res Int*. 2015; 2015: 128697.

Laboratory Talk. 'Analysis of benzoate and sorbate in soft drinks.' http://laboratorytalk.com/article/51192/analysis-of-benzoate-and-sorba 25 November 2003

Leishman, D., '"Original and Best"? How Barr's Irn-Bru Became a Scottish Icon.' *Études écossaises* [En ligne], 19 | 2017, mis en ligne le 01 avril 2017, consulté le 11 avril 2019. http://journals.openedition.org/etudesecoss-aises/1206

Lim, U., Subar, A. F., Traci, M., et al. 'Consumption of Aspartame-Containing Beverages and Incidence of Hematopoietic and Brain Malignancies.' *Cancer Epidemiol Biomarkers Prev.* 2006; 15(9): 1654–1659.

Live Science. 'Highly caffeinated drinks can impair cognitive abilities.' https://www.livescience.com/9081-highly-caffeinated-drinks-impair-cognitive-abilities.html 6 December 2010

Live Science. 'The truth about guarana.' https://www.livescience.com/36119-truth-guarana.html 27 January 2012

Lohner, S., Toews, I. & Meerpohl, J. J., 'Health outcomes of non-nutritive sweeteners: analysis of the research landscape.' *Nutr J.* 2017; 16(1): 55.

Lorjaroenphon, Y. & Cadwallader, K. R., 'Characterization of Typical Potent Odorants in Cola-Flavored Carbonated Beverages by Aroma Extract Dilution Analysis.' *J. Agric. Food Chem.* 2015; 63 (3): 769–775.

Ma, J., Fox, C. S., Jacques, P. F., et al. 'Sugar-sweetened beverage, diet soda, and fatty liver disease in the Framingham Heart Study cohorts.' *J Hepatol.* 2015; 63(2): 462–469.

Malik, V. S., 'Sugar sweetened beverages and cardiometabolic health.' *Curr Opin Cardiol.* September 2017; 32(5): 572–579.

McCann, D., et al. 'Food additives and hyperactive behaviour in 3-year-old and 8/9-year-old children in the community: a randomised, double-blinded, placebo-controlled trial.' *The Lancet.* 2007; 370(9598): 1560–1567.

Muraki, I., Imamura, F., Manson, J. E., et al. 'Fruit consumption and risk of type 2 diabetes: results from three prospective longitudinal cohort studies.' *BMJ.* 2013; 347: f5001.

Murphy, M. M., Barrett, E. C., Bresnahan, K. A., Barraj, L. M. '100% fruit juice and measures of glucose control and insulin sensitivity: a systematic review and meta-analysis of randomised controlled trials.' *J NutrSci.* 2017; 6:e59.

National Center for Complementary and Integrative Health. 'Energy drinks.' https://nccih.nih.gov/health/energy-drinks 26 July 2018

National Institute of Diabetes and Digestive and Kidney Diseases. 'Overweight and obesity statistics.' https://www.niddk.nih.gov/health-information/health-statistics/overweight-obesity August 2017

National Institutes of Health Office of Dietary Supplements. 'Potassium. Fact sheet for health professionals.' https://ods.od.nih.gov/factsheets/Potassium-HealthProfessional/ 5 March 2019

Natural Hydration Council. 'New research shows nationwide inappropriate use of sports drinks.' https://www.naturalhydrationcouncil.org.uk/press/new-research-shows-nationwide-inappropriate-use-of-sports-drinks/ 3 April 2012

Natural Hydration Council. 'Sports drinks fuel teens gaming and TV time.' https://www.naturalhydrationcouncil.org.uk/press/sports-drinks-fuel-teens-gaming-and-tv-time/

Neves, M. F., Trombin, V. G., Lopes, F. F., Kalaki, R. & Milan, P., 'Definition of juice, nectar and still drink.' *The orange juice business.* 2011. Wageningen Academic Publishers, Wageningen.

NHS. 'The truth about sweeteners.' https://www.nhs.uk/Livewell/Goodfood/Pages/the-truth-about-sucralose.aspx 28 February 2019

Nicoletti, M., 'Microalgae Nutraceuticals.' *Foods.* 2016; 5(3): 54.

Noakes, T. D., 'The role of hydration in health and exercise.' *BMJ.* 2012; 344: e4171.

Northwestern Extract. 'Introduction to the manufacture of soft drinks.' https://northwesternextract.com/manufacturing-of-soft-drinks/

Nursing Times. 'Sports drinks may have adverse effects on teens' dental health.' https://www.nursingtimes.net/news/news-topics/public-health/sports-drinks-may-have-adverse-effects-on-teens-dental-health/7006044.article 11 July 2016

Nyonya Cooking. 'Air mata kucing.' https://www.nyonyacooking.com/recipes/air-mata-kucing

Ocado. 'Rebel Kitchen Raw Organic Water.' https://www.ocado.com/webshop/product/Rebel-Kitchen-Raw-Organic-Coconut-Water/369010011

Ocado. 'Sibberi Bamboo Water Glow.' https://www.ocado.com/webshop/product/Sibberi-Bamboo-Water-Glow/349239011

Ocado. 'Sibberi Pure Birch Water.' https://www.ocado.com/webshop/product/Sibberi-Pure-Birch-Water/296930011

Open Food Facts. 'Vanilla Bean & Maple Syrup Smoothie – Marks & Spencer.' https://uk.openfoodfacts.org/product/00854467/vanilla-bean-maple-syrup-smoothie-marks-spencer 6 November 2014

OpenLearn. 'Fizzy drink.' http://www.open.edu/openlearn/science-maths-technology/science/chemistry/fizzy-drinks 26 September 2005

Peltier, S., Leprêtre, P. M., Metz, L. et al. 'Effects of Pre-exercise, Endurance, and Recovery Designer Sports Drinks on Performance During Tennis Tournament Simulation.' *Journal of Strength and Conditioning Research.* 2013; 27(11): 3076–3083.

Pérez-Idárraga, A. & Aragón-Vargas, L., 'Post-Exercise Rehydration with Coconut Water.' *Medicine and Science in Sports and Exercise.* 2010; 42.

Pound, C. M. & Blair, B., Canadian Paediatric Society. Nutrition and Gastroenterology Committee, Ottawa, Ontario, 'Energy and sports drinks in children and adolescents.' *Paediatrics & Child Health.* 2017; 22(7): 406–410.

PubChem. 'Dimethyl Dicarbonate (compound).' https://pubchem.ncbi.nlm.nih.gov/compound/3086#section=Pharmacology-and-Biochemistry

PubChem. 'Potassium Sorbate (compound).' https://pubchem.ncbi.nlm.nih.gov/compound/potassium_sorbate#section=Analytic-Laboratory-Methods

Richelsen, B., 'Sugar-sweetened beverages and cardio-metabolic disease risks.' *Current Opinion in Clinical Nutrition and Metabolic Care.* 2013; 16(4): 478–484.

Rogers, P. J. & Shahrokni, R., 'A Comparison of the Satiety Effects of a Fruit Smoothie, Its Fresh Fruit Equivalent and Other Drinks.' *Nutrients.* 2018; 10(4): 431.

Saat, M., Singh, R., Sirisinghe, R. G. & Nawawi, M., 'Rehydration after Exercise with Fresh Young Coconut Water, Carbohydrate-Electrolyte Beverage and Plain Water.' *Journal of Physiological Anthropology and Applied Human Science.* 2002; 21(2): 93–104.

Sainsbury's. 'Innocent Coconut Water.' https://www.sainsburys.co.uk/shop/gb/groceries/coconut-water-115152-44/innocent-coconut-water-1l

Sainsbury's Naked Coconut Water.' https://www.sainsburys.co.uk/shop/gb/groceries/naked-coconut-water-1l

Sainsbury's. 'Tymbark Cactus Drink.' https://www.sainsburys.co.uk/shop/gb/groceries/tymbark-cactus-drink-1l

Sainsbury's. 'Vita Coco Coconut.' https://www.sainsburys.co.uk/shop/gb/groceries/coconut-water-115152-44/vita-coco-100%25-pure-coconut-water-1l

Schimpl, F. C., da Silva, J. F., Gonçalves, J. F, & Mazzafera, P., 'Guarana: revisiting a highly caffeinated plant from the Amazon.' *J Ethnopharmacol.* 2013; 150(1): 14–31.

Schulze, M. B., Manson, J. E., Ludwig, D. S., et al. 'Sugar-Sweetened Beverages, Weight Gain, and Incidence of Type 2 Diabetes in Young and Middle-Aged Women.' *JAMA.* 2004; 292(8): 927–934.

Science Daily. 'Supertasters do not have particularly high density of taste buds on tongue, crowdsourcing says.' https://www.sciencedaily.com/releases/2014/05/140527161834.htm 27 May 2014

Sibberi. 'Tree water.' http://www.sibberi.com

Smithsonian.com. 'The benefits of probiotics might not be so clear cut.' https://www.smithsonianmag.com/science-nature/benefits-probiotics-might-not-be-so-clear-cut-180970221/ 6 September 2018

Snopes. 'Does Coca-Cola contain cocaine?' https://www.snopes.com/fact-check/cocaine-coca-cola/ 19 May 1999

Srinivasan, R., Smolinske, S. & Greenbaum, D., 'Probable gastrointestinal toxicity of Kombucha tea: is this beverage healthy or harmful?' *J Gen Intern Med*. 1997; 12(10): 643–644.

Steinman, H. A. & Weinberg, E. G., 'The effects of soft-drink preservatives on asthmatic children.' *S Afr Med J*. 1986; 70(7): 404–406.

Suez, J., Korem, T., Zilberman-Schapira, G., Segal, E. & Elinav, E., 'Non-caloric artificial sweeteners and the microbiome: findings and challenges.' *Gut Microbes*. 2015; 6(2): 149–155.

Sun, X., Ke, M. & Wang, Z., 'Clinical features and pathophysiology of belching disorders.' *Int J Clin Exp Med*. 2015; 8(11): 21906–21914.

Supply Chain. 'Drink it in. How your favorite soda is manufactured.' https://supplychainx.highjump.com/how-soda-is-manufactured.html 9 April 2018

Surjushe, A., Vasani, R. & Saple, D. G., 'Aloe vera: a short review.' *Indian J Dermatol*. 2008; 53(4):163–166.

Tappy, L. & Lê, K. A., 'Metabolic Effects of Fructose and the Worldwide Increase in Obesity.' *Physiological Reviews*. 2010; 90(1): 23–46.

Tayyem, R. F., Heath, D. D., Al-Delaimy, W. K. & Rock, C. L., 'Curcumin content of turmeric and curry powders.' *Nutr Cancer*. 2006; 55(2): 126–131.

Tesco. 'Tesco 100% Pure Squeezed Orange Juice With Bits.' https://www.tesco.com/groceries/en-GB/products/258997144

Tetra Pak. 'Juice, nectar and still drinks – easy to find your favourite.' https://www.tetrapak.com/findbyfood/juice-and-drinks/juice-nectar-still-drinks

The Atlantic. 'Is fermented tea making people feel enlightened because of . . . alcohol?' https://www.theatlantic.com/health/archive/2016/12/the-promises-of-kombucha/509786/ 8 December 2016

The Guardian. 'Birch water: the so-called superdrink you've never heard of.' https://www.theguardian.com/sustainable-business/2015/may/07/birch-water-so-called-superfood-superdrink-sustainability

The Guardian. 'Government to ban energy drink sales to children in England.' https://www.theguardian.com/business/2018/aug/29/ban-sale-energy-drinks-to-children-uk-government-combat-obesity 30 August 2018

The Guardian. 'How fruit juice went from health food to junk food.' https://www.theguardian.com/lifeandstyle/2014/jan/17/how-fruit-juice-health-food-junk-food 18 January 2014

The Guardian. 'Joint venture: Coca-Cola considers cannabis-infused range.'

https://www.theguardian.com/business/2018/sep/17/joint-venture-drinks-giant-coca-cola-mulls-cannabis-infused-range 18 September 2018

The New York Times. 'Dispute over Coca-Cola's secret formula.' https://www.nytimes.com/1993/05/03/business/dispute-over-coca-cola-s-secret-formula.html 3 May 1993

The Sugar Association. 'What is sugar?' https://www.sugar.org/sugar/what-is-sugar/

The UK Flavour Association. 'Flavourings are used to bring taste and variety to foods.' http://ukflavourassociation.org/about-us/what-are-flavourings

The Wall Street Journal. 'Wimbledon isn't just about tennis. There's also way too much squash.' https://www.wsj.com/articles/wimbledon-isnt-just-about-tennis-theres-also-way-too-much-squash-1467989111 8 July 2016

Thompson, M., Henegan, C. & Cohen, D., 'Food regulators must up their game.' *BMJ.* 2012; 345: e4753.

Tucker, K. L., Morita, K., Qiao, N., Hannan, M. T., Cupples, L. A. & Kiel, D. P., 'Colas, but not other carbonated beverages, are associated with low bone mineral density in older women: The Framingham Osteoporosis Study.' *Am J Clin Nutr.* 2006; 84(4): 936–942.

Unesda Soft Drinks Europe. 'Carbonated drink.' https://www.unesda.eu/lexikon/carbonated-drink/

Unesda Soft Drinks Europe. 'Preservatives.' https://www.unesda.eu/lexikon/preservatives/

Valdes, A. M., Walter, J., Segal, E. & Spector, T. D., 'Role of the gut microbiota in nutrition and health.' *BMJ.* 2018; 361: k2179.

Vally, H. & Misso, N. L., 'Adverse reactions to the sulphite additives.' *Gastroenterol Hepatol Bed Bench.* 2012; 5(1): 16–23.

Villarreal-Soto, S. A., Beaufort, S. & Bouajila, J., 'Understanding Kombucha Tea Fermentation: A Review.' *Concise Reviews & Hypotheses in Food Science.* 2018; 83(3): 580–588.

Vimto. 'Squash.' http://www.vimto.co.uk/squash.aspx#vimtoOriginal

Vina, I., Semjonovs, P., Linde, R. & Denina, I., 'Current Evidence on Physiological Activity and Expected Health Effects of Kombucha Fermented Beverage.' *Journal of Medicinal Food.* 2014; 17(2): 179–188.

Vogler, B. K. & Ernst, E., 'Aloe vera: a systematic review of its clinical effectiveness.' *Br J Gen Pract.* 1999; 49(447): 823–828.

Waitrose. 'Sibberi Maple Water.' https://www.waitrose.com/ecom/products/sibberi-maple-water/584584-519979-519980

Waitrose. 'Tapped Pure Birch Water.' https://www.waitrose.com/ecom/products/tapped-pure-birch-water/853930-601842-601843

Walters, D. E., 'Aspartame, a sweet-tasting dipeptide.' http://www.chm. bris.ac.uk/motm/aspartame/aspartameh.html February 2001

Woodward-Lopez, G., Kao, J. & Ritchie, L., 'To what extent have sweetened beverages contributed to the obesity epidemic?' *Public Health Nutrition.* 2011; 14(3): 499–509.

Yong. J. W. H., Ge, L., Ng, Y. F. & Tan, S. N., 'The Chemical Composition and Biological Properties of Coconut (*Cocos nucifera L.*) Water.' *Molecules.* 2009; 14(12): 5144–5164.

Zmora, N., et al. 'Personalized Gut Mucosal Colonization Resistance to Empiric Probiotics Is Associated with Unique Host and Microbiome Features.' *Cell.* 2018; 174(6): 1388–1405.e21

5. ALCOHOLIC DRINKS

Allen, A. L., McGeary, J. E. & Hayes, J. E., 'Polymorphisms in TRPV1 and TAS2Rs associate with sensations from sampled ethanol.' *Alcohol Clin Exp Res.* 2014; 38(10): 2550–2560.

ASCIA. 'Alcohol allergy.' https://www.allergy.org.au/patients/product-allergy/alcohol-allergy March 2019.

Ashurst, J. V. & Nappe, T. M., 'Methanol toxicity.' *Treasure Island (FL): StatPearls Publishing.* 15 March 2019.

Bais, S., Gill, N. S., Rana, N. & Shandil, S., 'A Phytopharmacological Review on a Medicinal Plant: *Juniperus communis.*' *Int Sch Res Notices.* 2014; 2014: 634723.

BBC. 'Beer before wine? It makes no difference to a hangover.' https://www.bbc.com/news/uk-47143368 8 February 2019

BBC. 'India toxic alcohol: At least 130 tea workers dead from bootleg drink.' https://www.bbc.com/news/world-asia-india-47341941 24 February 2019

Beer Store. 'What is beer?' http://www.thebeerstore.ca/beer-101

Bègue, L., Bushman, B. J., Zerhouni, O., Subra, B. & Ourabah, M. ' "Beauty is in the eye of the beer holder": People who think they are drunk also think they are attractive.' *British Journal of Psychology.* 2013; 104: 225–234.

CAMRA. 'What is real cider?' http://www.camra.org.uk/faqs

Choice. 'Preservatives in wine and beer.' https://www.choice.com.au/food-and-drink/drinks/alcohol/articles/preservatives-in-wine-and-beer 26 April 2016

Chow Hound. 'How are non-alcoholic beer and wine made?' https://www.chowhound.com/food-news/53912/how-are-nonalcoholic-beer-and-wine-made/ 4 April 2007

CNN. 'Toxic moonshine kills 102 in Mumbai slum.' https://edition.cnn.com/2015/06/22/asia/india-moonshine-deaths-mumbai/index.html 23 June 2015

Coeliac UK. 'Alcohol.' https://www.coeliac.org.uk/gluten-free-diet-and-lifestyle/keeping-healthy/alcohol/

Conscious Mixology. 'How are spirits made? (from seed to bottle)' http://www.consciousmixology.com/spirits-liqueurs-production/

de Gaetano, G., Costanzo, S., Di Castelnuovo, A. et al. 'Effects of moderate beer consumption on health and disease: A consensus document.' *Nutr Metab Cardiovasc Dis.* 2016; 26(6): 443–467.

Difford's Guide. 'Activated charcoal in cocktails.' https://www.diffords-guide.com/encyclopedia/1173/cocktails/activated-charcoal-in-cocktails

Difford's Guide. 'Does mixing drinks cause a worse hangover?' https://www.diffordsguide.com/encyclopedia/530/bws/does-mixing-drinks-cause-a-worse-hangover

Drinkaware. 'Low alcohol drinks.' https://www.drinkaware.co.uk/advice/how-to-reduce-your-drinking/how-to-cut-down/low-alcohol-drinks/

Drinkaware. 'Unit and calorie calculator.' https://www.drinkaware.co.uk/understand-your-drinking/unit-calculator

Drinks International. 'The world's best-selling classic cocktails 2018.' http://drinksint.com/news/fullstory.php/aid/7543/ 31 January 2018

Duffy, V. B., Davidson, A. C., Kidd, J. R., et al. 'Bitter receptor gene (TAS2R38), 6-n-propylthiouracil (PROP) bitterness and alcohol intake.' *Alcohol Clin Exp Res.* 2004; 28(11): 1629–1637.

Fever-Tree. 'The history of gin and tonic.' https://fever-tree.com/en_GB/article/gin-and-tonic-history

Francis Boulard & Fille. 'Champagne dosage.' https://www.francis-boulard.com/en/champagne-dosage.htm

Gizmodo. 'Happy hour: The science of non-alcoholic beer.' https://gizmodo.com/the-science-of-non-alcoholic-beer-509674407 25 May 2013

Goldberg, D. M., Hoffman, B., Yang, J. & Soleas, G. J., 'Phenolic Constituents, Furans, and Total Antioxidant Status of Distilled Spirits.' *J. Agric. Food Chem.* 1999; 47(10): 3978–3985.

Gorgus, E., Hittinger, M. & Schrenk, D., 'Estimates of Ethanol Exposure in Children from Food not Labeled as Alcohol-Containing.' *J Anal Toxicol.* 2016; 40(7): 537–542.

Gov.UK. 'Composition of foods integrated dataset (CoFID).' https://www.gov.uk/government/publications/composition-of-foods-integrated-dataset-cofid 25 March 2019

Gov.UK. 'New alcohol guidelines show increased risk of cancer.' https://www.gov.uk/government/news/new-alcohol-guidelines-show-increased-risk-of-cancer 8 January 2016

Griswold, M. G. et al. 'Alcohol use and burden for 195 countries and territories, 1990–2016: a systematic analysis for the Global Burden of Disease Study 2016.' *The Lancet.* 2018; 392(10152): 1015–1035.

Halsey, L. G., Huber, J. W., Bufton, R. D. J. & Little, A. C., 'An explanation for enhanced perceptions of attractiveness after alcohol consumption.' *Alcohol.* 2010; 44(4): 307–313.

Halsey, L. G., Huber, J. W. & Hardwick, J. C., 'Does alcohol consumption really affect asymmetry perception? A three-armed placebo-controlled experimental study.' *Addiction.* 2012; 107(7): 1273–1279.

Harvard Health Publishing. 'Ask the doctor: what causes red wine headaches?' https://www.health.harvard.edu/diseases-and-conditions/what-causes-red-wine-headaches

Harvard Health Publishing. 'Is red wine actually good for your heart?' https://www.health.harvard.edu/blog/is-red-wine-good-actually-for-your-heart-2018021913285 19 February 2018

Harvard Health Publishing. 'Will tonic water prevent nighttime leg cramps?' https://www.health.harvard.edu/bone-and-muscle-health/will-tonic-water-prevent-nighttime-leg-cramps September 2016

Harvard T. H. Chan. 'Study says no amount of alcohol is safe, but expert not convinced.' https://www.hsph.harvard.edu/news/hsph-in-the-news/alcohol-risks-benefits-health/ 2018

Haseeb, S., Alexander, B. & Baranchuk, A., 'Wine and Cardiovascular Health A Comprehensive Review.' *Circulation.* 2017; 136: 1434–1448.

Höferl, M., Stoilova, I., Schmidt, E., et al. 'Chemical Composition and Antioxidant Properties of Juniper Berry (*Juniperus communis L.*) Essential Oil. Action of the Essential Oil on the Antioxidant Protection of Saccharomyces cerevisiae Model Organism.' *Antioxidants* (Basel). 2014; 3(1): 81–98.

Moreno-Indias, I., 'Benefits of the beer polyphenols on the gut microbiota.' *Nutr Hosp.* 2017; 15 (34(Suppl 4)): 41–44.

Jensen, W. B., 'The Origin of Alcohol "Proof".' *J. Chem. Educ.* 2004; 81: 1258.

Laurel Gray. '5 stages of the wine making process.' http://laurelgray.com/5-stages-wine-making-process/ 14 November 2014

LiveScience. 'Traces of the world's first "microbrew" found in a cave in Israel.' https://www.livescience.com/63631-oldest-beer-brewing-evidence.html 20 September 2018

Kinnek. 'Pot still vs. column still: what's the difference?' https://www.kinnek.com/article/pot-still-vs-column-still-whats-the-difference/#/ 12 May 2016

Mackus, M., Adams, S., Barzilay, A., et al. 'Proceeding of the 8th Alcohol Hangover Research Group Meeting.' *Curr Drug Abuse Rev.* 2017; 9(2): 106–112.

Maintz, L. & Novak, N., 'Histamine and histamine intolerance.' *Am J Clin Nutr.* 2007; 85(5): 1185–1196.

Market Research World. 'Cool down for alcopops.' http://www.marketresearchworld.net/content/view/370/77/14 November 2005

Martini. 'Martini meets Rossi.' https://www.martini.com/uk/en/we-are-martini/

Meister, K. A., Whelan, E. M. & Kava, R., 'The Health Effects of Moderate Alcohol Intake in Humans: An Epidemiologic Review.' *Critical Reviews in Clinical Laboratory Sciences.* 2000; 37(3): 261–296.

Metro. 'Gin fans – you've been making martinis all wrong.' https://metro.co.uk/2016/02/18/gin-fans-youve-been-making-martinis-all-wrong-5703089/ 18 February 2016

Munchies. 'This is why teenagers aren't drinking alcopops anymore.' https://munchies.vice.com/en_us/article/8qkd74/this-is-why-teenagers-arent-drinking-alcopops-anymore 20 October 2015

National Institute on Alcohol Abuse and Alcoholism. 'Alcohol metabolism: an update.' https://pubs.niaaa.nih.gov/publications/aa72/aa72.htm July 2007

News.com.au. 'Trio died drinking $2 moonshine so lethal one sip could paralyse drinkers' arms for 15 minutes.' https://www.news.com.au/national/courts-law/trio-died-drinking-2-moonshine-so-lethal-one-sip-could-paralyse-drinkers-arms-for-15-minutes/news-story/ 23 November 2016

NHS. 'Alcohol support.' https://www.nhs.uk/live-well/alcohol-support/calculating-alcohol-units/ 13 April 2018

NHS. 'Beer and bone strength.' https://www.nhs.uk/news/food-and-diet/beer-and-bone-strength/ 5 March 2009

NHS. 'Is a pint of beer a day good for the heart?' https://www.nhs.uk/news/heart-and-lungs/is-a-pint-of-beer-a-day-good-for-the-heart/ 12 May 2016

NHS. 'Moderate drinking may reduce heart disease risk.' https://www.nhs.uk/news/heart-and-lungs/moderate-drinking-may-reduce-heart-disease-risk/ 23 March 2017

NHS Choices. 'What's your poison? A sober analysis of alcohol and health in the media. A Behind the Headlines special report.' October 2011

Oxford Living Dictionary. 'What is the origin of the phrase "hair of the dog"?' https://en.oxforddictionaries.com/explore/what-is-the-origin-of-the-phrase-hair-of-the-dog/

Penning, R., van Nuland, M., Fliervoet, L. A. L., Olivier, B. & Verster, J. C., 'The Pathology of Alcohol Hangover.' *Current Drug Abuse Reviews.* 2010; 3(2): 68–75.

Phobia Wiki. 'Dipsophobia.' http://phobia.wikia.com/wiki/Dipsophobia

Phobia Wiki. 'Methyphobia.' http://phobia.wikia.com/wiki/Methyphobia

Phobia Wiki. 'Zythophobia.' http://phobia.wikia.com/wiki/Zythophobia

Piasecki, T. M., Robertson, B. M. & Epler, A. J., 'Hangover and risk for alcohol use disorders: existing evidence and potential mechanisms.' *Curr Drug Abuse Rev.* 2010; 3(2): 92–102.

Pittler, M. H., Verster, J. C. & Ernst, E., 'Sex, Drugs, And Rock And Roll: Interventions for preventing or treating alcohol hangover: systematic review of randomised controlled trials.' *BMJ.* 2005; 331: 1515.

Prat, G., Adan, A. & Sánchez-Turet, M., 'Alcohol hangover: a critical review of explanatory factors.' *Hum Psychopharmacol.* 2009; 24(4): 259–267.

Rohsenow, D. J., Howland, J., Arnedt, J. T., Almeida, A. B., Greece, J., Minsky, S., Kempler, C. S. & Sales, S., 'Intoxication With Bourbon Versus Vodka: Effects on Hangover, Sleep, and Next-Day Neurocognitive Performance in Young Adults.' *Alcoholism: Clinical and Experimental Research.* 2010; 34: 509–518.

Rohsenow, D. J. & Howland, J., 'The role of beverage congeners in hangover and other residual effects of alcohol intoxication: a review.' *Curr Drug Abuse Rev.* 2010; 3(2): 76–79.

Schirone, M., Visciano, P., Tofalo, R. & Suzzi, G., 'Histamine Food Poisoning.' *Handb Exp Pharmacol.* 2017; 241: 217–235.

Smithsonian.com. 'The deadly side of moonshine.' https://www.smithsonianmag.com/smart-news/the-deadly-side-of-moonshine-41629081/ 18 September 2012

Stevenson, C., 'Hans Off!: The Struggle for Hans Island and the Potential Ramifications for International Border Disupte Resolution.' *Boston College International and Comparative Law Review.* 2007; 30(1 – Article 16): 263–275.

Stockwell, T., Zhao, J., Panwar, S., Roemer, A., Naimi, T. & Chikritzhs, T., 'Do "Moderate" Drinkers Have Reduced Mortality Risk? A Systematic Review and Meta-Analysis of Alcohol Consumption and All-Cause Mortality.' *J Stud Alcohol Drugs.* 2016; 77(2): 185–198.

The Alcohol Free Shop. 'Frequently asked questions (FAQs).' https://www.alcoholfree.co.uk/faqs.php

The Australian Wine Research Institute. 'Fining agents.' https://www.awri. com.au/industry_support/winemaking_resources/frequently_asked_ questions/fining_agents/

The Conversation. 'Is mixing drinks actually bad?' https://theconversation. com/is-mixing-drinks-actually-bad-87256 29 December 2017

The Guardian. 'Notes and queries: James Bond requested that his Martini be "shaken not stirred" – would it make any difference?' https://www. theguardian.com/notesandqueries/query/0,,-2866,00.html

The New York Times. 'Canada and Denmark fight over island with whisky and schnapps.' https://www.nytimes.com/2016/11/08/world/what-in-the-world/canada-denmark-hans-island-whisky-schnapps.html 7 November 2016

The Telegraph. 'Gin sales triple as Brits turn to high-end booze.' https:// www.telegraph.co.uk/news/2018/07/03/gin-sales-triple-brits-turn-high-end-booze/ 3 July 2018

The Telegraph. 'Is the alcopop back in fashion?' https://www.telegraph. co.uk/finance/newsbysector/retailandconsumer/11399498/Is-the-alco-pop-back-in-fashion.html 8 February 2015

The Wine Cellar Insider. 'How to produce and make red or white wine explained.' https://www.thewinecellarinsider.com/wine-topics/wine-educational-questions/how-wine-is-made/

Topiwala, A., Allan, C. L. & Valkanova, V., 'Moderate alcohol consumption as risk factor for adverse brain outcomes and cognitive decline: longitudinal cohort study.' *BMJ.* 2017, 357: j2353.

TrendHunter Lifestyle. 'Wellness cocktail. Superfood cocktails combine flavors of the moment from both worlds.' https://www.trendhunter. com/protrends/wellness-cocktail

Trevithick, C. C., Chartrand, M. M., Wahlman, J., Rahman, F., Hirst, M. & Trevithick, J. R., 'Shaken, not stirred: bioanalytical study of the antioxidant activities of martinis.' *BMJ.* 1999; 319(7225): 1600–1602.

U.S. News. 'The 6 healthiest cocktail ingredients.' https://health.usnews. com/health-news/blogs/eat-run/articles/2017-06-30/the-6-healthiest-cocktail-ingredients 30 June 2017

Vally, H. & Thompson, P. J., 'Allergic and asthmatic reactions to alcoholic drinks.' *Addict Biol.* 2003; 8(1): 3–11.

Vassilopoulou, E., Karathanos, A. & Siragakis, G., et al. 'Risk of allergic reactions to wine, in milk, egg and fish-allergic patients.' *Clin Transl Allergy.* 2011; 1(1): 10.

Verster, J. C. & Penning, R., 'Treatment and prevention of alcohol hangover.' *Curr Drug Abuse Rev.* 2010; 3(2): 103–109.

Verster, J.C., Stephens, R., Penning, R., et al. 'The alcohol hangover research group consensus statement on best practice in alcohol hangover research.' *Curr Drug Abuse Rev.* 2010; 3(2): 116–126.

VinePair. 'All the ways to make champagne and sparkling wine, explained.' https://vinepair.com/articles/sparkling-wine-champagne-methods/ 26 November 2017

VinePair. 'How distilling works.' https://vinepair.com/spirits-101/how-distilling-works/

VinePair. 'The 10 most popular beer brands in the world.' https://vinepair.com/articles/10-biggest-beer-brands-world-2017/ 11 September 2017

VOA News. '100 deaths highlight Indonesia's bootleg booze problem.' https://www.voanews.com/a/indonesia-deaths-illegal-alcohol/4346422.html 13 April 2018

Wantke, F., Götz, M. & Jarisch, R., 'The red wine provocation test: intolerance to histamine as a model for food intolerance.' *Allergy Proc.* 1994; 15(1): 27–32.

Weiskirchen, S. & Weiskirchen, R., 'Resveratrol: How Much Wine Do You Have to Drink to Stay Healthy?' *Adv Nutr.* 2016; 7(4): 706–718.

Wine Folly. 'How sparkling wine is made.' https://winefolly.com/review/how-sparkling-wine-is-made/

Wine From Here. 'Sulfur dioxide (SO2) in wine).' https://winobrothers.com/2011/10/11/sulfur-dioxide-so2-in-wine/ 11 October 2011

Wine Guy. 'Destemming grapes.' http://www.wineguy.co.nz/index.php/81-all-about-wine/920-destemming-grapes 10 September 2017

World Atlas. 'Hans Off! Canada and Denmark's arctic dispute.' https://www.worldatlas.com/articles/hans-island-boundary-dispute-canada-denmark-territorial-conflict.html 25 April 2017

World Cancer Research Fund. 'Alcoholic drinks and the risk of cancer.' https://www.wcrf.org/dietandcancer/exposures/alcoholic-drinks 2018

World Health Organization. 'Global status report on alcohol and health 2018.' http://apps.who.int/iris/bitstream/handle/10665/274603/9789241565639-eng.pdf?ua=1

DIGESTIF

ABC News. 'Nightline report: "When grassroots protest rallies have corporate sponsors".' https://abcnews.go.com/Nightline/video/grassroots-protest-rallies-corporate-sponsors-26671038

Alcohol Change UK. 'Alcohol industry influence on public policy: A case study of pricing and promotions policy in the UK.' https://alcoholchange.org.uk/publication/alcohol-industry-influence-on-public-policy-a-case-study-of-pricing-and-promotions-policy-in-the-uk 20 September 2012

Bragg, M. A., Miller, A. N., Elizee, J., Dighe, S. & Elbel, B. D., 'Popular Music Celebrity Endorsements in Food and Nonalcoholic Beverage Marketing.' *Pediatrics*. 2016; 138(1)e20153977.

Forbes. 'As U.S. soda sales fizzle, Coca-Cola and PepsiCo target developing nations.' https://www.forbes.com/sites/nancyhuehnergarth/2016/02/09/as-u-s-soda-sales-fizzle-coca-cola-and-pepsico-target-developing-nations/#7d213a111cec 9 February 2016

Godlee, F., 'Minimum alcohol pricing: a shameful episode.' *BMJ*. 2014; 348: g110.

Gornall, J., 'Under the influence: Scotland's battle over alcohol pricing.' *BMJ*. 2014; 348: g1274.

Hawkins, B., Holden, C. & McCambridge, J., 'Alcohol industry influence on UK alcohol policy: A new research agenda for public health.' *Crit Public Health*. 2012; 22(3): 297–305.

Hollywood Branded. 'Top celebrity beverage endorsers.' https://blog.hollywoodbranded.com/top-celebrity-beverage-endorsements 6 February 2018

Investopedia. 'Much of the global beverage industry is controlled by Coca Cola and Pepsi.' https://www.investopedia.com/ask/answers/060415/how-much-global-beverage-industry-controlled-coca-cola-and-pepsi.asp 14 November 2018

Investopedia. 'A look at Coca-Cola's advertising expenses.' https://www.investopedia.com/articles/markets/081315/look-cocacolas-advertising-expenses.asp 6 October 2018

Marketing Schools. 'Celebrity marketing.' http://www.marketing-schools.org/types-of-marketing/celebrity-marketing.html

Social Media Week. 'Celebrity endorsements on social media are driving sales and winning over fans.' https://socialmediaweek.org/blog/2015/09/brands-using-celebrity-endorsements/ 30 September 2015

Thacker, P., 'Coca-Cola's secret influence on medical and science journalists.' *BMJ*. 2017; 357: j1638.

The BMJ. 'Alcohol pricing.' https://www.bmj.com/content/alcohol-pricing

The Drinks Business. 'Top 10 celebrity drinks launches of 2017.' https://www.thedrinksbusiness.com/2017/12/top-10-celebrity-drinks-launches-of-2017/2/ 20 December 2017

The Drinks Business. 'Supreme court backs minimum alcohol pricing in Scotland.' https://www.thedrinksbusiness.com/2017/11/supreme-court-backs-minimum-alcohol-pricing-in-scotland/ 15 November 2017

The Guardian. 'Coca-Cola and other soft drinks firms hit back at sugar tax plan.' https://www.theguardian.com/business/2016/mar/17/coca-cola-hits-back-at-sugar-tax-plan 18 March 2016

The Hollywood Reporter. 'Meet the Hollywood "Brandfather" who's pairing Aaron Rodgers with beef jerky'. https://www.hollywoodreporter.com/news/meet-hollywood-brandfather-whos-pairing-811758 29 July 2015

The Telegraph. 'Coca-Cola "spends millions on research to prove that fizzy drinks don't make you fat".' https://www.telegraph.co.uk/finance/news-bysector/retailandconsumer/11920984/Coca-Cola-spends-millions-on-research-to-prove-that-fizzy-drinks-dont-make-you-fat.html 9 October 2015

The Telegraph. 'Were ministers under the influence of drinks industry?' https://www.telegraph.co.uk/news/politics/10557347/Were-ministers-under-the-influence-of-drinks-industry.html 7 January 2014

The Washington Post. 'How business funded the anti-soda tax coalition.' https://www.washingtonpost.com/news/monkey-cage/wp/2014/11/24/how-business-funded-the-anti-soda-tax-coalition/ 24 November 2014

Union of Concerned Scientists. 'How Coca-Cola disguised its influence on science about sugar and health.' https://www.ucsusa.org/disguising-corporate-influence-science-about-sugar-and-health#.XB09qy2cZ3k

Index

A2 milk 49–50, 61–2
absinthe 207
acesulfame potassium 145, 146, 160
acetaldehyde 52, 216
acetate 216
acetic acid 175
acidity regulators 149, 151
activated carbon 7
activated charcoal 182–4, 213
Adams, Douglas xii
adrenaline 128
Advocaat 207
aeropress 119
air mata kucing 178–9
airag 52
Alcohol by Volume (ABV) 190, 191,
     192, 200
  alcopops 214
  beer 192, 194
  cider 198
  spirits and liqueurs 207
  wine 192
alcohol dehydrogenase (ADH) 216
alcoholic drinks 188–227
  alcohol phobias 227
  alcoholic intake, genetics and 191,
     216
  allergic reactions to 224–6
  breastfeeding and 43–4
  calorie content 220–1
  combined with energy drinks 167–8
  daily fluid intake and xiv
  defined 189–90
  global consumption of 189
  'hair of the dog' 223
  hangovers 184, 220, 221–4, 226
  health benefits and harms 215–20
  metabolism of xiii, 215–16

minimum alcohol pricing proposals
     234
  mixing drinks 226
  national associations 189
  nutrients in 221
  'proof' 190–1
  units 190
  see also alcopops; beer; cider; cocktails;
     liqueurs; perry; spirits; wine
alcopops 214
aldehyde dehydrogenase (ALDH) 216
ales 194, 195
alkaline water 3, 17–18
allergic reactions 41, 45, 61, 75, 152,
     154, 224–6
almond milk 66, 68, 70–1, 77
aloe vera juice 169–70
Amaretto 207
Americano 121, 123
amino acids 39, 73, 94, 100, 104, 135,
     166
Angostura bitters 211, 212
anti-ageing properties 17, 20, 182
anti-inflammatory action 24, 58, 63,
     83, 101, 109, 113, 123, 132, 177, 178,
     219
antibiotics 50
  in milk 65–6
  in water 31
antioxidant properties 24, 99, 101, 103,
     112, 122, 132, 170, 175, 177, 178,
     214, 218–19, 220
appetite regulation 127
aquifers 2, 14
arsenic 76
artesian water 14
artificial sweeteners 144–7, 160
aspartame 145, 146–7

aspartic acid 146
athletes xiv, 20, 22, 126–7
  see also sports drinks
atole 81

B vitamins 22, 47, 57, 58, 74, 135, 170,
  177
baby milks 36–46
  breast milk 37–8, 41–2, 43, 60, 79,
    237
  formula milks 38–46
bacteria 3, 4, 7, 13, 15, 16, 17, 26, 29,
    30, 31, 42, 48, 56–7, 63, 65, 100,
    110, 153, 175, 179, 199
Bailey's Irish Cream 207
bamboo water 173–4
barley water 140, 159, 160
beer 154, 193–8, 214, 221, 224
  ABV 194
  brewing and fermentation 193, 194–5
  carbonation 193, 194
  'gluten-free'/'gluten-removed' 225
  health effects 219–20
  low- and non-alcoholic versions 192–3
  types of 194–6
'beer goggles' 196–8
benzene 153
benzoates 153
biofilms 30, 110
birch water 172–3
bisphenol A (BPA) 28–9
bitter lemon 215
bitters (beer) 195
black tea 92, 93, 94, 101, 102, 107–9
blood pressure regulation 104, 109,
  132
blood sugar levels 135, 142, 143, 156
blue-green algae 182
bock beers 196
bone health 53, 90, 127–8, 151, 173, 219
bottled waters 1, 7, 11–25, 31, 32
  alkaline water 3, 17–18
  artesian water 14
  distilled water 13–14
  enhanced waters 19–21

global consumption of 11–12
mineral water and spring water 12–13
plastic bottles 1, 28–30, 31
purified water 13
raw water 14–16
rosemary water 23–5
sparkling water 18–19
vitamin waters 22–3
world record 32
bouillon drinks 81
bourbon 211, 224, 230
brain freeze 27–8
brandy 208, 209
breast milk 37–8, 60, 79, 237
  and formula milks compared 41–6
  health benefits 41–2, 43
  nutrients 37
bubble tea 115
buffalo milk 47
burping 139
buttermilk 52, 158

Ca Phe Trung 122
Café de Olla 122
cafestol 123
Caffe Latte 121
Caffe Mocha 121
caffeine
  daily intake of 84–5, 90
  decaffeination process 86–8, 89
  in energy drinks 83, 84, 85, 165, 166,
    167, 168
  health benefits and harms 83, 89–90,
    103–4
  in hot drinks 83–5, 89–90, 100, 101,
    103–4, 114, 117, 122, 126–7, 132
  military personnel and 168
  and physical performance 126–7
  in soft drinks 150, 151
  stimulatory effects 83, 90, 100
calcium 4, 5, 6, 13, 37, 54–5, 74, 170
camel milk 47, 52
Campari 207
cancer
  anti-cancer effects 24, 41, 53, 63, 101,

102, 103, 104, 109, 123, 125, 132, 177, 219
cancer risk 22, 29, 53, 64, 75–6, 88, 114–15, 136, 146, 217
cannabidiol (CBD) 185
Cappuccino 121
carbohydrates xiii, 37, 72, 141
carbon dioxide 3, 13, 18, 19, 29, 87, 151, 190, 199
carbonated soft drinks 139, 141, 161, 174
  carbonation 151, 161
carbonated water see sparkling water
carbonation
  beer 193, 194
  soft drinks 151, 161
  water 18–19
carcinogens 88, 153
  see also cancer
carrageenan 77
casein 39, 40, 50, 61, 102, 135
cashew milk 70
catechins 99, 101, 103, 109, 110, 111
celebrity endorsements vii, 113, 229 31
cereal milks 69, 71, 72, 73, 75
chai 108
chamomile tisane 113
Champagne 204, 205, 225–6, 231
chlorine 3, 7, 8
chlorogenic acids (CGAs) 122–3, 128
cholesterol 53–4, 63, 72, 73, 103, 108, 123
chromogens 124
cider 154, 198–9, 221
  cider-making process 198–9
  types of 199
citric acid 151
club soda 215
Coca-Cola 23, 84, 144, 171, 185, 232–3
  formula 150
cockroaches 124
cocktails 211–14, 221
cocoa 128–33
  bitterness 132
  cultivation and processing 129–30

health benefits 131–3
hot chocolate 130
hot cocoa 130
in malted drinks 134
nutrients 131–2
roasting process 129
coconut milk 66, 70, 71, 72, 73
coconut water 171–2, 213, 231
coffee 115–28, 230
  bitterness 117, 122
  brewing 118–21
  caffeine 84, 85, 117, 122, 126–7, 128
  constituents of 122–4
  cultivation and processing 116–18
  decaffeination process 86–8
  global consumption of 116
  global production of 116
  health benefits and harms 82–3, 122–8
  instant coffee 118
  'joe' 120
  optimal daily amount 128
  roasting process 87–8, 89, 117, 123
coffee equipment and styles
  aeropress 119
  Americano 121, 123
  Ca Phe Trung 122
  Café de Olla 122
  cafetière coffee 119, 123
  Caffe Latte 121
  Caffe Mocha 121
  Cappuccino 121
  coffee pods and capsules 121–2
  espresso 119, 121
  filter coffee 119
  Flat White 121
  Irish coffee 122
  Machiatto 121
  Mazagran 122
  Pharisäer coffee 122
  Pour Over 119
  stove-top espresso makers 119
  Turkish coffee 122, 123
  Yuanyang (Kopi Cham) 122
cognac 220

cognitive performance 24, 41, 43, 103, 104, 109, 125–6, 168
Cointreau 207
colas 83, 144, 151, 215
cold drinks (non-alcoholic) 139–87
  human preference for 140
  see also soft drinks
colostrum 37
colourings 152
condensed milk 51
congeners 209, 220, 224
contaminants
  in milks 43–5, 57
  in water 3–4, 7, 8, 16
cordials *see* liqueurs; squashes and cordials
cow milk 46, 47–50
  A2 milk 49–50, 61–2
  antibiotics in 65–6
  condensed and evaporated milks 51
  dairy industry, environmental impacts of 66, 77

dandelion and burdock 140
decaffeinated drinks 84, 85–9, 127
  decaffeination process 86–8, 89
  health issues 88–9
dehydration xiii, 21
dementia 103, 104, 125, 126
dental health 9–11, 18, 43, 109–11, 124, 142, 160, 164
detox teas 113–14
detoxing, concept of 20, 114, 183, 184–5
diabetes 24, 41, 43, 58, 61, 106, 125, 142, 144, 158, 182
dimethyl dicarbonate 153, 159–60
disaccharides 141
distillation 208–9
distilled water 13–14
diterpenes 123, 128
drinking water 1–32
drinks
  definition of xii
  metabolism of xii–xiii

fat content 50–1, 53–4, 73
fermented milks 52, 63–5
filtered milk 48
fortified milk 47, 54
free school milk 35
health benefits and harms 35–6, 52–66
homogenisation 48, 58–9
lactose intolerance 49, 59–61
lactose-free milk 60
milk allergy 41, 45, 61, 75
night-time milk 67
nutrients 47, 74
organic 50, 55
pasteurisation 48, 57
processing 47–9
promotion of 35, 36
raw milk 47–8, 56–8
skimmed/semi-skimmed/whole/full-cream 50–1, 54, 72
UHT milk 48–9, 58
curcumin 178

drinks industry *see* marketing
drugs
  in breast milk 44
  in water 30–1
dunkels 196

E numbers 149, 152
electrolytes 20–1
energy drinks 1, 83, 84, 85, 127, 161, 165–8, 177
  health benefits and harms 166–7
enhanced waters 19–21
  added electrolytes 20–1
  oxygen(ated) water 20
espresso 119, 121
esters 148, 209
ethanol 175, 190, 191, 207, 208, 215–16, 224, 226
ethyl acetate (EA) 87, 88, 89
evaporated milk 51
excretion xiii, xiv

exercise 20, 21, 22, 163, 164
  see also sports drinks

fermentation 190
  beer 193, 194–5
  cider 198–9
  malolactic 199, 202
  secondary 204
  wine 201, 202, 204, 207
fermented milks 51–2, 63–5
fibre, dietary 72, 131, 181
filmjolk 52
fining agents 224–5
fizzy drinks *see* carbonated soft drinks
Flat White 121
flavonoids 103, 132, 218–19
flavourings in soft drinks 148–9, 159
flax milk 69, 70
fluid loss xiii, 21, 22, 163
fluoride and fluoridation 9–11, 13, 15,
  104–5
fluorosis 9, 10, 105
formula milks 38–46
  and breast milk compared 41–6
  casein-dominant milks 39
  follow-on milk 39–40
  goat-milk-based 41
  goodnight milks 40
  growing-up milk/toddler milks 40
  nutrients 39
  soya-based milks 40–1
  specialised formula 40
  whey-dominant milks 39, 40
fructose 141, 142–3, 156
fruit juices 155–8
  'freshly squeezed' 157
  fruit 'comminute' 157
  fruit concentrate 156–7
  fruit nectar 155
  health harms 156, 158
  legislative requirements 155
  'not from concentrate' 157
  nutrients 156
  pasteurisation 157
  reconstituted juice 157

smoothies 157, 158
still drinks 155
sugar in 156, 157
water in 156
functional drinks *see* energy drinks;
    sports drinks; wellness drinks

gastrointestinal problems 16, 19, 27,
  77
ghrelin 19
gin 207–8, 209, 211, 212, 220
ginger ale 149, 215
glucose 141, 142, 143, 156
glucuronolactone 166
gluten 75, 225
GM foods 78
goat milk 41, 47, 52, 61
goodnight milks 40
Grand Marnier 207
green tea 92, 93, 94, 96, 99, 103, 108,
  109–11, 114
  health benefits and harms 101,
    109–11
  matcha 111
guar 154
guarana 83, 165–6
Guinness 195
gut microbiome 42, 123, 145, 146, 175,
  179–81

'hair of the dog' 223
hangovers 184, 220, 221–4, 226
Hans Island 188–9
hard water 4–6, 94
hazelnut milk 72
heart health 6, 22, 53, 58, 59, 61, 73,
  108, 109, 125, 132, 219
helles lagers 196
hemp milk 69, 70, 72, 73
herbal infusions and tisanes 112–14
herbal products, breastfeeding and 44
high fructose corn syrup (HFCS)
  142–3
histamines 225–6
homogenisation of milk 48, 58–9, 70

hops 194
Horlicks 134
hormones in milk 66
horse and donkey milk 47, 52
hot chocolate 130–1
hot drinks 81–138
    caffeine 83–5, 89–90, 100, 101,
        103–4, 114, 117, 122, 126–7, 132
    decaffeinated drinks 84, 85–9
    in hot weather 136
    potential harms 82, 136
    self-heating cans 137–8
    see also cocoa; coffee; malted milk
        drinks; tea
hyperactivity 152, 153, 167
hypertension 90, 132
hypertonic drinks 163
hyponatraemia xiii–xiv
hypotonic drinks 163

ice wine 206–7
ice-cold drinks 25–8
ice-cream headache 27–8
iced coffee and tea 59, 115, 122
immune system functioning 37, 42,
        57, 63, 175, 182
Indian Pale Ale (IPA) 195
insulin production 135, 142
iodine 55, 74
ion-exchange 5–6
ionisation 17, 18
ionised alkaline water see alkaline
        water
Irish coffee 122
Irn-Bru 162
iron 3, 12, 40, 43, 102, 103, 109, 162,
        170
isoflavones 75
isotonic drinks 163

jasmine tea 109
juice drinks see fruit juices

Kahlua 207
kahweol 123

kefir 52, 63–4, 179
kombucha 174–6, 213
Kopi Cham 122
kule naoto 52
kumis 52

L-theanine 99–100, 101, 104
lactic acid 52, 199
lactic acid bacteria 65
lactic acidosis 176
lactose 37, 47, 52, 60, 75, 141
lactose intolerance 49, 59–61, 75
lactose-free milk 60
lagers 194–5, 196
lambic beers 195
lassi 158
legume-based milks 68–9
    see also soya milks
lemonade 149, 215
Limoncello 207
liqueurs 207
locust bean gum 154
longan 178
LUXE milkshake 80

McDonald's coffee 82–3
Machiatto 121
magnesium 4, 5, 6, 13, 37, 135, 170
malic acid 151, 199
malted milk drinks 133–6
maltose 141
manganese 12, 173
maple water 173
marijuana 69, 185
marketing drinks 229–37
    celebrity endorsements xii, 113,
        229–31
    multinational companies 232–5
    political lobbying 233, 234–5
    research funding 233–4, 235
martinis 212, 213, 214
matcha tea 111
Mazagran 122
melatonin 67, 135
methanol 146, 209, 210

methylene chloride (MC) 87, 88, 89
methylxanthines 83, 103–4, 122, 128, 132
microalgae 182
microplastics 31
mild (beer) 195
milks 33–80
  animal milks *see* milks, animal
  baby milks 36–46
  in hot drinks 96–7, 100, 102, 124, 135
  LUXE milkshake 80
  milk-related records 33
  plant-based milks 33–4, 36, 66–79
milks, animal 33, 34–5, 46–66, 79, 237
  buttermilk 52, 158
  composition and nutrients 46–7, 79
  fermented milks 51–2, 63–5
  global consumption of 34
  health benefits and harms 35–6, 52–66
  water in 47, 72, 77
  World Milk Day 36
  see also cow milk, *and other specific entries*
Milo 134
mineral water 12–13
mixers 215
monk fruit 178
monosaccharides 141
moonshine 210
mursik 64

niacin (vitamin B3) 123, 170
night-time milk 67
nut milks 66, 68, 70–1, 72, 74, 75, 77

oat milk 69, 71, 72, 73, 75
obesity 41, 43, 142, 143
  childhood 53, 144, 164–5
office tea run 100
oolong tea 93, 94, 114
oral rehydration solution 21
organic milks 50, 55
organic wines 204, 225
Ouzo 207

Ovaltine 134
oxygen(ated) water 20

pale ales 195
palm oil 46
Pan Galactic Gargle Blaster xii
pasteurisation
  fruit juices 157
  milks 48, 57, 70
pea milk 69, 71, 73
pectin 154
peppermint tisane 113
perry 198
pH 3, 8, 17, 18, 151
Pharisäer coffee 122
pharmaceuticals
  in breast milk 44
  in water 30–1
phenylalanine 146
phosphoric acid 151
pilsners 196
plant-based milks 66–79
  calorie content 72, 73
  cereal milks 69, 71, 72, 73, 75
  choosing 78–9
  environmental impacts 77–8
  fat content 73
  fortified milks 74
  genetically modified raw material 78
  growing popularity of 66–8
  health benefits and harms 71–7
  legume-based milks 68–9
  nut milks 66, 68, 70–1, 72, 74, 75, 77
  nutritional content 72–5
  processing 70
  seed milks 69–70
  sugar in 72–3
  taste 70–1
  water in 72, 77
  see also specific entries
plastic bottles 28–30
polyethylene terephthalate (PET)-bottled waters 29
polyphenols 94, 97, 101, 102, 110, 112, 132, 175, 218–19, 220

polysaccharides 141
polystyrene cups 95–6
port 207, 209
porters 195
potable water *see* drinking water
potassium 13, 21, 37, 170, 172, 221
potassium sorbate 153, 159
prebiotics 39, 123, 179, 181
pregnant women 25, 55, 56, 176
  caffeine and 90, 127
preservatives 152–4, 159–60
prickly pear juice 170–1
probiotics 15, 39, 42, 56, 57, 64, 65,
    175, 179–81
proteins xiii, 37–8, 39, 40, 43, 47, 50,
    61, 70, 73, 97, 102
  whey protein drinks 62–3
Pu-erh tea 93, 94
purified water 13
pyridines 123

quercetin 218–19
quinine 215

rainwater 4
raw milk 56–8
raw water 14–16
rice milk 69, 70, 72, 73, 76–7
risk-seeking behaviours 167, 168
Robinsons Barley Water 160
rooibos 83, 112–13
root beer 140
rosemary water 23–5
rum 207, 208, 209, 211, 214

saccharin 145, 147
sap waters 172–4
saturated fats 53, 54, 73
schwarzbiers 196
scientific studies
  cherry picking 186–7
  funding 233–4, 235
  interpretation of xv–xvi
scoby 174, 175
seed milks 69–70

self-heating cans 137–8
serotonin 135
sesame milk 72, 75
sheep milk 47
sherry 207, 209
silica 173–4
sleep-inducing effects 67, 134–5
smoothies 157, 158
social media influencers 237
sodas *see* carbonated soft drinks
sodium xiii, 5, 6, 13, 21, 22, 37
sodium benzoate 153
soft drinks 139–87
  acidity regulators '49 151
  carbonated *see* carbonated soft drinks
  celebrity endorsements 230
  colourings 152
  definition of 140
  dilutable drinks 158–9
  flavoured waters 159–60
  flavourings 148–9, 159
  fruit juices *see* fruit juices
  functional drinks *see* energy drinks;
    sports drinks; wellness drinks
  history of 139–40
  ingredients and health-related
    concerns 141–55
  mixers 215
  preservatives 152–4, 159–60
  stabilisers 154–5
  still and juice drinks 155–8
  sugar and artificial sweeteners 141–7
  water in 141, 159–60, 161
Soft Drinks Industry Levy (UK)
    143–4
soft water 4, 5, 6, 94, 141
sorbates 153
soya milks 66, 68–9, 70, 73, 74, 75
  environmental impacts 77–8
  formula milks 40–1
  GM soya 78
  health effects 69, 75–6
sparkling water 18–19
sparkling wines 204–5
spelt milk 69

spirits 207–10, 220
  see also specific entries
spirulina 182
sports drinks 161, 163–5, 172, 229, 231
  celebrity endorsements 231
  constituents of 163
  hypertonic drinks 163
  hypotonic drinks 163
  isotonic drinks 163
  weight gain and 164, 165
sports performance 20, 103, 126–7,
    128, 163
spring water 12–13
squashes and cordials 158–9
stabilisers 154–5
sterilisation of milks 70
steviol glycosides 145, 147, 160
stouts 195
sucralose 145, 160
sucrose 141, 142, 143, 156
sugar
  in alcoholic drinks 221
  in condensed milk 51
  in energy drinks 165, 166, 167
  in hot drinks 134–5
  in plant milks 72–3
  in soft drinks 141–4, 156, 157, 163,
    171–2
  sugar substitutes *see* artificial
    sweeteners
sugar tax 143–4, 233
sujeonggwa 179
sulphites 153–4, 203–4, 225, 226
sunflower milk 69
supertasters 148, 191

tannins 94, 95, 99, 101, 102, 124, 199,
    201, 209, 226
tap water 2–11, 13, 15, 21
taurine 166
tea 91–115
  beneficial ingredients/health effects
    101–7, 108, 109–10
  bitterness 94, 95, 96, 102
  black tea 92, 93, 94, 101, 102, 107–9

caffeine 84, 85, 94, 101, 103–4, 114
  components of 101–5
  crush-tear-curl (CTC) processing 93,
    99
  cultivation and processing 92–3, 106
  daily consumption, limits on 108–9
  decaffeination process 88
  global consumption of 91
  green tea 92, 93, 94, 96, 99, 101, 103,
    108, 109–11, 114
  herbal infusions and tisanes 112–14
  iced tea and bubble tea 115
  kombucha 174–6
  microwave tea-making 96
  milk in tea 96–7, 100, 102
  office tea run 100
  oolong tea 93, 94, 114
  oxidation 92, 93
  in polystyrene cups 95–6
  Pu-erh tea 93, 94
  tannins 94, 95, 99, 101, 102
  tea making 93–5, 96
  teabags 97–100
  teaware 95, 96–7
  varieties of 91–2
  white tea 92–3, 94, 110
  yerba mate 83, 114–15
teeth staining 102, 124
tequila 208, 209, 212, 231
theobromine 103, 132
theophylline 103, 114
thirst xiv
tisanes 112–14
tonic water 215
toxins xiii, 183, 184
  see also detoxing, concept of
triglycerides 143
trigonelline 123–4
tryptophan 135
Turkish coffee 122, 123
turmeric 177–8

UHT milk 48–9, 58
urine xiii, xiv

Vimto 159
vitamin A 54, 170
vitamin C 22, 47, 57, 58, 153, 170, 178
vitamin D 54, 57, 74
vitamin E 55, 74, 170
vitamin waters 22–3
vodka 207, 208, 209, 212, 214, 224, 231

water filter jugs 7–8
water intoxication xiii–xiv
water softeners 5–6
waters 1–32, 237
  in animal milks 47, 77
  in the body xiii
  bottled *see* bottled waters
  chemical formula 2
  contamination 3–4, 7, 8, 16, 26–7, 30–1
  daily water needs xiv
  drinking too little/too much xiii–xiv
  drinking water 1–32
  flavoured water 159–60
  in fruit juices 156
  hard water 3, 4–6, 94
  ice 25–8
  in milks 47, 72, 77
  pharmaceuticals in 30–1
  rainwater 4
  in soft drinks 141, 159–60, 161, 171–4
  soft water 3, 4, 5, 6, 94, 141
  sources 2, 12
  tap water 2–11, 13, 15, 21
  in tea 94
  treatment 2–3, 4, 5–6, 7, 7–8, 9–11, 12–14
weight loss 19, 103, 104, 110, 113, 144
wellness drinks xii, 161, 168–87
  active ingredients 176–85
  cherry-picked scientific studies 186–7
  cocktails 212–13

dubious health claims 236–7
  marketing 168–9, 230–1, 236–7
  plant-based drinks 169–76
wheat beers 195–6
whey protein drinks 62–3
whisky 207, 209, 211, 214, 220, 230
  moonshine 210
white tea 92–3, 94, 110
wine 154, 200–7
  ABV 200
  dessert wines 206
  fining/filtration 203
  fortified wines 200, 207
  health effects 218–19
  ice wine 206–7
  low- and non-alcoholic versions 192–3
  organic wines 204, 225
  red wines 201, 202, 205, 206, 218, 221, 224, 225, 226
  rosé wines 201, 206
  sparkling wines 204–5
  vegan wine 203
  white wines 201–2, 205–6, 221
  winemaking 200–5
World Milk Day 36
world records 32, 33, 80

xanthan 154
xanthine oxidase (XO) 58–9

yak milk 47, 52
Yakult 52
yeasts 63, 153, 175, 179, 190, 194–5, 196, 198–9, 201, 202
yellow tea 91
yerba mate 83, 114–15
yogurt drinks 64–5, 179
Yuanyang 122

zoonosis infections 56